SpringerBriefs in Speech Technology

Series Editor:
Amy Neustein

For other titles published in this series, go to
http://www.springer.com/series/10043

Editor's Note

The authors of this series have been hand-selected. They comprise some of the most outstanding scientists – drawn from academia and private industry – whose research is marked by its novelty, applicability, and practicality in providing broad based speech solutions. The SpringerBriefs in Speech Technology series provides the latest findings in speech technology gleaned from comprehensive literature reviews and *empirical investigations* that are performed in both laboratory and *real life* settings. Some of the topics covered in this series include the presentation of real life commercial deployment of spoken dialog systems, contemporary methods of speech parameterization, developments in information security for automated speech, forensic speaker recognition, use of sophisticated speech analytics in call centers, and an exploration of new methods of soft computing for improving human-computer interaction. Those in academia, the private sector, the self service industry, law enforcement, and government intelligence, are among the principal audience for this series, which is designed to serve as an important and essential reference guide for speech developers, system designers, speech engineers, linguists and others. In particular, a major audience of readers will consist of researchers and technical experts in the automated call center industry where speech processing is a key component to the functioning of customer care contact centers.

Amy Neustein, Ph.D., serves as Editor-in-Chief of the International Journal of Speech Technology (Springer). She edited the recently published book "Advances in Speech Recognition: Mobile Environments, Call Centers and Clinics" (Springer 2010), and serves as quest columnist on speech processing for Womensenews. Dr. Neustein is Founder and CEO of Linguistic Technology Systems, a NJ-based think tank for intelligent design of advanced natural language based emotion-detection software to improve human response in monitoring recorded conversations of terror suspects and helpline calls. Dr. Neustein's work appears in the peer review literature and in industry and mass media publications. Her academic books, which cover a range of political, social and legal topics, have been cited in the Chronicles of Higher Education, and have won her a pro Humanitate Literary Award. She serves on the visiting faculty of the National Judicial College and as a plenary speaker at conferences in artificial intelligence and computing. Dr. Neustein is a member of MIR (machine intelligence research) Labs, which does advanced work in computer technology to assist underdeveloped countries in improving their ability to cope with famine, disease/illness, and political and social affliction. She is a founding member of the New York City Speech Processing Consortium, a newly formed group of NY-based companies, publishing houses, and researchers dedicated to advancing speech technology research and development.

David Suendermann

Advances in Commercial Deployment of Spoken Dialog Systems

 Springer

David Suendermann
SpeechCycle, Inc.
26 Broadway 11th Floor
New York, NY 10004
USA
david@speechcycle.com

ISSN 2191-737X   e-ISSN 2191-7388
ISBN 978-1-4419-9609-1   e-ISBN 978-1-4419-9610-7
DOI 10.1007/978-1-4419-9610-7
Springer New York Dordrecht Heidelberg London

Library of Congress Control Number: 2011930670

Printed on acid-free paper

Springer is part of Springer Science+Business Media (www.springer.com)

Preface

Spoken dialog systems have been the object of intensive research interest over the past two decades, and hundreds of scientific articles as well as a handful of text books such as [25, 52, 74, 79, 80, 83] have seen the light of day. What most of these publications lack, however, is a link to the "real world", i.e., to conditions, issues, and environmental characteristics of deployed systems that process millions of calls every week resulting in millions of dollars of cost savings. Instead of learning about:

- Voice user interface design.
- Psychological foundations of human-machine interaction.
- The deep academic[1] side of spoken dialog system research.
- Toy examples.
- Simulated users.

the present book investigates:

- Large deployed systems with thousands of activities whose calls often exceed 20 min of duration.
- Technological advances in deployed dialog systems (such as reinforcement learning, massive use of statistical language models and classifiers, self-adaptation, etc.).
- To which extent academic approaches (such as statistical spoken language understanding or dialog management) are applicable to deployed systems – if at all.

---

[1] This book draws a line between core research on spoken dialog systems as performed in academic institutions and in large industrial research labs on the one hand and commercially deployed spoken dialog systems on the other hand. As a convention, the former will be referred to as *academic*, the latter as *deployed* systems.

To Whom It May Concern

There are three main statements touched upon above:

1. Huge commercial significance of deployed spoken dialog systems.
2. Lack of scientific publications on deployed spoken dialog systems.
3. Overwhelming difference between academic and deployed systems.

These arguments, further backed up in Chap. 1, indicate a strong need for a comprehensive overview about the state of the art in deployed spoken dialog systems. Accordingly, major topics covered by the present book are as follows:

- After a brief introduction to the general architecture of a spoken dialog system, Chap. 1 offers some insight into important parameters of deployed systems (such as traffic, costs) before *comparing the worlds of academic and deployed spoken dialog systems* in various dimensions.
- *Architectural paradigms* for all the components of deployed spoken dialog systems are discussed in Chap. 2. This chapter will also deal with the many limitations deployed systems face (with respect to e.g. functionality, openness of input/output language, performance) imposed by hardware requirements, legal constraints, and the performance and robustness of current speech recognition and understanding technology.
- The key to success or failure of deployed spoken dialog systems is their performance. Performance being a diffuse term when it comes to the (continuous) *evaluation of dialog systems*, Chap. 3 will be dedicated to why, what, and when to measure performance of deployed systems.
- After setting the stage for a continuous performance evaluation, the logical consequence is trying to increase system performance on an ongoing basis. This attempt is often realized as a continuous cycle involving multiple *techniques for adapting and optimizing* all the components of deployed spoken dialog systems as discussed in Chap. 4. Adaptation and optimization are essential to deployed applications because of two main reasons:

  1. Every application can only be suboptimal when deployed for the first time due to the absence of live data during the initial design phase. Hence, application tuning is crucial to make sure deployed spoken dialog systems achieve maximum performance.
  2. Caller behavior, call reasons, caller characteristics, and business objectives are subject to change over time. External events that can be of irregular (such as network outages, promotions, political events), seasonal (college football season, winter recess), or slowly progressing nature (slow migration from analog to digital television, expansion of the Smartphone market) may have considerable effects on what type of calls an application must be able to handle.

Due to the book's focus on paradigms, processes, and techniques applied to deployed spoken dialog systems, it will be of primary interest to speech scientists, voice user interface designers, application engineers, and other technical staff of the automated call center industry, probably the largest group of professionals in the speech and language processing industry. Since Chap. 1 as well as several other parts of the book aim at bridging the gap between academic and deployed spoken dialog systems, the community of academic researchers in the field is in focus as well.

New York City   *David Suendermann*
February 2011

Acknowledgements

The name of the series which the present book is a volume of, SpringerBriefs, makes use of two words that have a meaning in the German language: Springer (knight) and Brief (letter). Indeed, I was fighting hard like a knight to get this letter done in less than four months of sleepless nights. In this effort, several remarkable people stood by me: Dr. Amy Neustein, Series Editor of the SpringerBriefs in Speech Technology, whose strong editing capabilities I learned to greatly appreciate in a recent similar project, kindly invited me to author the present monograph. Essential guidance and support in the course of this knight ride came also from the editorial team at Springer – Alex Greene and Andrew Leigh. On the final spurt, Dr. Roberto Pieraccini as well as Dr. Renko Geffarth contributed invaluable reviews of the entire volume adding the finishing touches to the manuscript.

Contents

Chapter 1
Deployed vs. Academic Spoken Dialog Systems

Abstract After a brief introduction into the architecture of spoken dialog systems, important factors of deployed systems (such as call volume, operating costs, or induced savings) will be reviewed. The chapter also discusses major differences between academic and commercially deployed systems.

Keywords Academic dialog systems • Architecture • Call automation • Call centers • Call traffic • Deployed dialog systems • Erlang-B formula • Operating costs and savings

1.1 At-a-Glance

Spoken dialog systems are today the most massively used applications of speech and language technology and, at the same time, the most complex ones. They are based on a variety of different disciplines of spoken language processing research including:

- Speech recognition [25].
- Spoken language understanding [75].
- Voice user interface design [22].
- Spoken language generation [111].
- Speech synthesis [129].

As shown in Fig. 1.1, generally, a spoken dialog system receives input speech from a conventional telephony or Voice-over-IP switch and triggers a speech recognizer whose recognition hypothesis is semantically interpreted by the spoken language understanding component. The semantic interpretation is passed to the dialog manager hosting the system logic and communicating with arbitrary types of backend services such as databases, web services, or file servers. Now, the dialog manager generates a response generally corresponding to one or more pre-defined

D. Suendermann, *Advances in Commercial Deployment of Spoken Dialog Systems,*
SpringerBriefs in Speech Technology, DOI 10.1007/978-1-4419-9610-7_1,
© Springer Science+Business Media, LLC 2011

Fig. 1.1 General diagram of a spoken dialog system

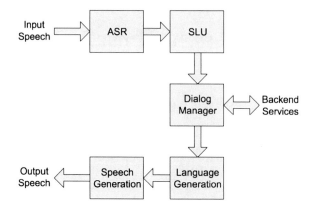

semantic symbols that are transformed into a word string by the language generation component. Finally, a text-to-speech module transforms the word string into audible speech that is sent back to the switch[1].

1.2   Census, Internet, and a Lot of Numbers

In 2000, the U.S. Census counted 281,421,906 people living in the United States [1]. The same year, the Federal Communication Commission reported that common telephony carriers handled 537 billion local calls that amount to over 5 daily calls per capita on average [3]. While the majority of these calls were of a private nature, a huge number were directed to customer care contact centers (aka call centers) often serving as the main communication channel between a business and its customers. Although over the past 10 years, Internet penetration has grown enormously (traffic has increased by factor 224 [4]) and, accordingly, many customer care transactions are carried out online, the amount of call center transactions of large businesses is still extremely large.

For example, a large North-American telecommunications provider serving a customer base of over 5 million people received more than 40 million calls into its service hotline in the time frame October 2009 through September 2010 [confidential source]. Considering that the average duration (aka handling time) of the processed calls was about 8 min, the overall access minutes of this period $(326 \cdot 10^6 \text{ min})$ can be divided by the duration of the period (346 days = 525,600 min) to calculate the average number of concurrent calls. For the present example, it is 621.

---

[1]See Sect. 2.5 for differences in language and speech generation between academic and deployed spoken dialog systems.

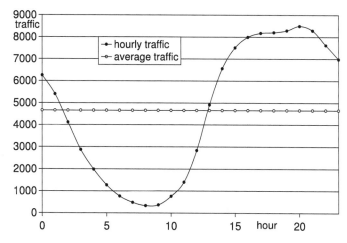

Fig. 1.2 Distribution of call traffic into the customer service hotline of a large telecommunication provider

Does this mean, 621 call center agents are required all year round? No, this would be considerably underestimated bearing in mind that traffic is not evenly distributed throughout the day and the year.

Figure 1.2 shows the distribution of hourly traffic over the day for the above mentioned service hotline averaged over the time period October 2009 through September 2010. It also displays the average hourly traffic which is about 4,700 calls. The curve reaches a minimum of 334 calls, i.e. only the 15th part of the average, at 8AM UTC. Taking into account that the telecommunication company's customers are located in the four time zones of the contiguous United States and that they also observe daylight saving time, the time lap between UTC and the callers' time zone varies between 4 and 8 h. In other words, minimum traffic is expected sometime between 12 and 4AM depending on the actual location. On the other hand, the curve's peak is at 8PM UTC (12 to 4PM local time) with about 8,500 received calls which is a little less than twice the average.

Apparently, it would be an easy solution to scale call center staff according to the hours of the day, i.e., less people at night, more people in peak hours. Unfortunately, in the real world, the load is not as evenly distributed as suggested by the averaged distribution of Fig. 1.2. This is due to a number of reasons including:

- Irregular events of predictable (such as promotion campaigns, roll-outs of new products) or unpredictable nature (weather conditions, power outages).
- Regular/seasonal events (e.g., annual tax declaration, holidays), but also
- The randomness of when calls come in:
  Consider the above mentioned minimum average hourly volume of $n = 334$ calls and an average call length of 8 min. Now, one can estimate the probability that k calls overlap as

$$p_k(n, p) = \binom{n}{k} p^k (1-p)^{(n-k)} \tag{1.1}$$

with $p = 8$ min/60 min. Equation (1.1) is the probability mass function of a binomial distribution. If you had m call center agents, the probability that they will be enough to handle all incoming traffic is

$$P_m(n, p) = \sum_{k=0}^{m} p_k(n, p) = I_{1-p}(n - m, m + 1) \qquad (1.2)$$

with the regularized incomplete beta function I [5]. P_m is smaller than 1 for $m < n$, i.e., there is always a chance that agents will not be able to handle all traffic unless there are as many agents as the total number of calls coming in, simply because, theoretically, all calls could come in at the very same time. However, the likelihood that this happens is very small and can be controlled by (1.2), which, by the way, can also be derived using the Erlang-B formula, a widely used statistical description of load in telephony switching equipment [77]. For example, to make sure that call center agents are capable of handling all incoming traffic in 99% of the cases, one would estimate

$$\hat{m} = \arg\min_m |P_m(n, p) - 0.99|. \qquad (1.3)$$

For the above values for n and p, one can compute $\hat{m} = 60$. On the other hand, simply averaging traffic as

$$\bar{m} = np \qquad (1.4)$$

(which is the expected value of the binomial distribution) produces $\bar{m} = 44.5$. Consequently, even if the average statistics of Fig. 1.2 would hold true, 45 agents at 8 AM GMT would certainly not suffice. Instead, 60 agents would be necessary to cover 99% of traffic situations without backlog. Figure 1.3 shows how the ratio between $\hat{m}$ and $\bar{m}$ evolves for different amounts of traffic given the above defined p. The higher the traffic, the closer the ratio gets to the theoretical 1.0 where as many agents are required as suggested by the averaged load.

In addition to the expected unbalanced load of traffic, the above listed irregular and regular/seasonal events lead to a significantly higher variation of the load. To get a more comprehensive picture of this variation, every hour's traffic throughout the collection period was measured individually and displayed in Fig. 1.4 in order of decreasing load.

This graph (with a logarithmic abscissa) shows that, over more than 15% of the time, traffic was higher than twice the average (displayed as a dashed line in Fig. 1.4) and that there were occasions when traffic exceeded the quadruple average. Again, assuming that e.g. 99% of the situations (including exceptional ones) are to be handled without backlog, one would still need to handle situations of up to 12,800 incoming calls per hour producing $\hat{m} = 1,797$.

This number shows that there would have to be several thousand call center agents available to deal with this traffic unless efficient automated self-service solutions are deployed to complement the task of human agents. Call center

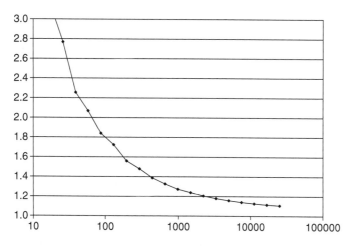

Fig. 1.3 Ratio between $\bar{m}$ and $\hat{m}$ depending on the number of calls per hour with $p = 8$ min/60 min and 99% coverage without backlog

Fig. 1.4 Hourly call traffic into the customer service hotline of a large telecommunication provider measured over a period of one year in descending order

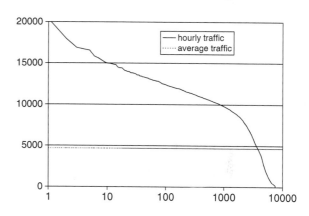

automation by means of spoken dialog systems thus can bring very large savings considering that [10]:

1. The average cost to recruit and train per agent is between $8,000 and $12,000.
2. Inbound centers have an average annual turnover of 26%.
3. The average hourly wage median is $15.

Assuming a gross number of 3,000 agents for the above customer, (1) would produce some $24M to $36M just for the initial agent recruiting and training. (2) and (3) combined would produce a yearly additional expense of almost $90M if the whole traffic would be handled entirely by human agents.

In contrast, if certain (sub-)tasks of the agent loop would be carried out by automated spoken dialog systems, costs could be significantly reduced. Once a spoken dialog system is built, it is easily scalable just by rolling out the respective piece of software on additional servers. Consequently, (1) and (2) are minimal. The operating costs of a deployed spoken dialog system including hosting, licensing, or telephony fees would usually be in the range of a few cents per minute, drastically reducing the hourly expense projected by (3). These considerations highly support the use of automated spoken dialog systems to take over certain tasks in the realm of the business of customer contact centers such as, for instance:

- Call routing [141]
- Billing [38]
- FAQ [30]
- Orders/sales [40]
- Hours, branch, department, and product search [20]

Table 1.1 Major differences between academic and deployed spoken dialog systems

	Area	Academic systems	Deployed systems	Further reading
1	Speech recognition	Statistical language models	Rule-based grammars, few statistical language models	Sections 2.3.1 and 2.3.2
2	Spoken language understanding	Statistical named entity tagging, semantic tagging, (shallow) parsing [9, 78, 87]	Rule-based grammars, key-word spotting, few statistical classifiers [54, 120, 128]	Sections 2.3.1 and 2.3.2
3	Dialog management	MDP, POMDP, inference [63, 66, 143]	Call flow, form-filling [86, 89, 108]	Section 2.4
4	Language generation	Statistical, rule-based	Manually written prompts	Section 2.5
5	Speech generation	Text-to-speech synthesis	Pre-recorded prompts	Section 2.5
6	Interfaces	Proprietary	VoiceXML, SRGS, MRCP, ECMAScript [19, 32, 47, 72]	Sections 2.6 and 2.3.1
7	Data and technology	Often published and open source	Proprietary and confidential	
8	Typical dialog duration	40 s, 5 turns [29]	277 s, 10 turns [confidential source]	
9	Corpus size	100s of dialogs, 1000s of utterances [29]	1,000,000s of dialogs and utterances [118]	
10	Typical applications	Tourist information, flight booking, bus information [28, 65, 96]	Call routing, package tracking, phone billing, phone banking, technical support [6, 43, 76, 88]	
11	Number of scientific publications	Many	Few	

- Directory assistance [108]
- Order/package tracking [107]
- Technical support [6] or
- Surveys [112].

1.3   The Two Worlds

For over a decade, spoken dialog systems have proven their effectiveness in commercial deployments automating billions of phone transactions [142]. For a much longer period of time, *academic* research has focused on spoken dialog systems as well [90]. Hundreds of scientific publications on this subject are produced every year, the vast majority of which originate from academic research groups.

As an example, at the recently held Annual Conference of the International Speech Communication Association, Interspeech 2010, only about 10% of the publications on spoken dialog systems came from people working on deployed systems. The remaining 90% experimented with:

- Simulated users, e.g. [21, 55, 91, 92].
- Conversations recorded using recruited subjects, e.g. [12, 49, 62, 69], or
- Corpora available from standard sources such as the Linguistic Data Consortium (LDC) or the Spoken Dialog Challenge, e.g. [97].

Now, the question arises on how and to which extent the considerable endeavor of the academic research community affects what is actually happening in deployed systems. In an attempt to answer this question, Table 1.1 compares academic and deployed systems along multiple dimensions specifically reviewing the five main components shown in Fig. 1.1. It becomes obvious that *differences* dominate the picture.

Chapter 2
Paradigms for Deployed Spoken Dialog Systems

Abstract This chapter covers state-of-the-art paradigms for all the components of deployed spoken dialog systems. With a focus on speech recognition and understanding components as well as dialog management, the specific requirements of deployed systems will be discussed. This includes their robustness against distorted and unexpected user input, their real-time-ability, and the need for standardized interfaces.

Keywords Components of spoken dialog systems • Confirmation • Dialog management • Language generation • Natural language call routing • Real-time systems • Rejection • Robustness • Rule-based grammars • Speech recognition • Speech understanding • Speech synthesis • Statistical classifiers • Statistical language models • Voice browsing • VoiceXML

2.1 A Few Remarks on History

After half a century of intensive research into automatic speech recognition (one of the first published functional speech recognizers was built at Bell Labs in 1952 [27]), in the 1990s, the technology finally achieved a performance (in terms of accuracy and speed) that could be applied to simple tasks in the telephony systems of companies with large customer care call volume. Solutions to phone-based self-service using touch-tone interaction already existed. Now, applications could be speech-enabled allowing for a much wider range of solutions helping companies like FedEx, American Airlines, or UPS to effectively expand their self-service customer support offerings [88]. Applications ranged from package tracking (with a tracking number specified by the caller) to stock quotes and flight schedule information.

D. Suendermann, *Advances in Commercial Deployment of Spoken Dialog Systems*,
SpringerBriefs in Speech Technology, DOI 10.1007/978-1-4419-9610-7_2,
© Springer Science+Business Media, LLC 2011

Speech-enabled menus have clear advantages compared to touch-tone menus when it comes to:

- Input items distinguishing a large number of types (such as city names or stock listings) or
- Mixed initiative or over-specification – when spoken language understanding and dialog manager are designed accordingly, the caller can input information or formulate requests unexpected at the current point of the dialog, e.g.

> S: Where would you like to depart from?
> C: From JFK on January 5th.

Another milestone in the development of spoken dialog systems was the introduction of *natural language call routing* in 1997 by researchers at AT&T Labs [43]. Large companies (as AT&T itself) often have a single service hotline phone number that serves as a portal to a variety of services and call center departments with different skills. There can be dozens if not hundreds of different destinations conventionally requiring a multi-layered menu structure to get the caller to the right place as shown in the following dialog recorded from the service hotline of a cable provider:

> S: Which one can I help you with: Your bill, tech support, an order, an appointment, or a change to your service?
> C: Repeat.
> S: Just say: *My bill*, or press 1. *Tech support*, or press 2. *An order*, or press 3. *An appointment*, or press 4. Or say *make a change to my service*, or press 5.
> C: Tech support.
> S: And which one are you having problems with: Internet, e-mail, cable TV, or telephone?
> C: My telephone.
> S: Are you having trouble with the dial tone on your phone?
> C: Nay.
> S: Then, do you have a question about a feature on your phone such as voicemail or call forwarding?
> C: Yeah.
> S: You can say *voice mail*, *caller ID*, or *call forwarding*. If it is not one of those, say *help me with something else*.
> C: Repeat the message, please.
> S: Just say: *Voice mail*, or press 1. *Caller ID*, or press 2. *Call forwarding*, or press 3. Or, if it is not one of those, say *help me with something else*, or press 4.
> C: Help me with something else.
> S: Here are some others to choose from: Call waiting, three-way calling, speed dialing, call return, anonymous call rejection, or custom ring. Again, if it is not one of these, say *help me with something else*.
> C: Three-way calling.
> S: I can explain you how three-way calling works or help you with problems with it. Just say *instructions* or *problems*. Or you can say *help me with something else*.
> C: Problems.

In contrast, a conversation using a natural language call router with an open-ended question could have been

> S: Briefly tell me what you are calling about today.
> C: My three-way calling is not working.

Apparently, this type of prompting comes along with a much shorter handling time resulting in a number of substantial advantages:

- Handling fees are saved (considering the processing of millions of such calls, the shaving of just seconds for every call can result in a significant impact on the application's bottom line).
- By reducing the number of recognition events necessary to get a caller to the right place, the chance of recognition errors decreases as well (even though it is true that open-ended question contexts perform worse than directed dialog, e.g., 85% vs. 95% True Total[1], the fact that doing several of the latter in a row exponentially decreases the chance that the whole conversation completes without error – e.g. the estimated probability that five user turns get completed without error is $(95\%)^5 = 77\%$ which is already way lower than the performance of the open-ended scenario; for further reading on measuring performance, see Chap. 3). Reducing recognition errors raises the chance of automating the call without intervention of a human agent.
- User experience is also positively influenced by shortening handling time, reducing recognition errors, and conveying a smarter behavior of the application [35].
- Open-ended prompting also prevents problems with callers not understanding the options in the menu and choosing the wrong one resulting in potential misroutings.

The underlying principle of natural language call routing is the automatic mapping of a user utterance to a finite number of well-defined classes (aka categories, slots, keys, tags, symptoms, call reasons, routing points, or buckets). For instance, the above utterance

> My three-way calling is not working

was classified as Phone_3WayCalling_Broken, in a natural language call routing application distinguishing more than 250 classes [115]. If user utterances are too vague or out of the application's scope, additional directed disambiguation questions may be asked to finally route the call. Further details on the specifics of speech recognition and understanding paradigms used in deployed spoken dialog systems are given in Sect. 2.3.

2.2   Components of Spoken Dialog Systems

As introduced in Sect. 1.1 and depicted in Fig. 1.1, spoken dialog systems consist of a number of components (speech recognition and understanding, dialog manager, language and speech generation). In the following sections, each of

---

[1] See Sect. 3.2 for the definition of this metric.

these components will be discussed in more detail focusing on deployed solutions and drawing brief comparisons to techniques primarily used in academic research to date.

2.3 Speech Recognition and Understanding

In Sect. 2.1, the use of speech recognition and understanding in place of the formerly common touch-tone technology was motivated. This section gives an overview about techniques primarily used in deployed systems as of today.

2.3.1 Rule-Based Grammars

In order to commercialize speech recognition and understanding technology for their application in dialog systems, at the turn of the millennium, companies such as Sun Microsystems, SpeechWorks, and Nuance made the concept of *speech recognition grammar* popular among developers. Grammars are essentially a specification "of the words and patterns of words to be listened for by a speech recognizer" [47,128]. By restricting the scope of what the speech recognizer "listens for" to a small number of phrases, two main issues of speech recognition and understanding technology at that time could be tackled:

1. Before, large-vocabulary speech recognizers had to recognize every possible phrase, every possible combination of words. Likewise, the speech understanding component had to deal with arbitrary textual input. This produced a significant margin of error unacceptable for commercial applications. By constraining the recognizer with a small number of possible phrases, the possibility of errors could be greatly reduced, assuming that the grammar covers all of the possible caller inputs. Furthermore, each of the possible phrases in a grammar could be uniquely and directly associated with a predefined semantic symbol, thereby providing a straightforward implementation of the spoken language understanding component.
2. The strong restriction of the recognizer's scope as well as the straightforward implementation of the spoken language understanding component significantly reduced the required computational load. This allowed speech servers to process multiple speech recognition and understanding operations simultaneously. Modern high-end servers can individually process more than 20 audio inputs at once [2].

Similar to the industrial standardization endeavor on VoiceXML described in Sect. 2.6, speech recognition grammars often follow the W3C Recommendation SRGS (Speech Recognition Grammar Specification) published in 2004 [47].

2.3.2   Statistical Language Models and Classifiers

Typical contexts for the use of rule-based grammars are those where caller responses are highly constrained by the prompt such as:

- Yes/No questions (*Are you calling because you lost your Internet connection?*).
- Directed dialog (*Which one best describes your problem: No picture, missing channels, error message, bad audio...?*).
- Listable items (city names, phone directory, etc.).
- Combinatorial items (phone numbers, monetary amounts, etc.).

On the other hand, there are situations where rule-based grammars prove impractical because of the large variety of user inputs. Especially, responses to open prompts tend to vary extensively. For example, the problem collection of a cable TV troubleshooting application uses the following prompt:

Briefly tell me the problem you are having in one short sentence.

The total number of individual collected utterances of this context was so large that the rule-based grammar resulting from the entire data used almost 100 MB memory which proves unwieldy in production server environments with hundreds of recognition contexts and dozens of concurrent calls. In such situations, the use of statistical language models and classifiers (statistical grammars) is recommendable. By generally treating an open prompt such as the one above as a call routing problem (see Sect. 2.1), every input utterance is associated with exactly one class (the routing point). For instance, responses to the above open prompt and their associated classes are:

Um, the Korean channel doesn't work well ∘—• Channel_Other
The signal is breaking up ∘—• Picture_PoorQuality
Can't see HBO ∘—• Channel_Missing
My remote control is not working ∘—• Remote_NotWorking
Want to purchase pay-per-view ∘—• Order_PayPerView_Other

This type of mapping is generally produced semi-automatically as further discussed in Sect. 4.1.

The utterance data can be used to train a statistical language model that is applied at runtime by the speech recognizer to generate a recognition hypothesis [100]. Both the utterances and the associated classes can be used to train statistical classifiers that are applied at runtime to map the recognition hypothesis to a semantic hypothesis (class). An overview about state-of-the-art classifiers used for spoken language understanding in dialog systems can be found in [36].

The initial reason to come up with the rule-based grammar paradigm was that of avoiding too complex search trees common in large-vocabulary continuous speech recognition (see Sect. 2.3.1). This makes the introduction of statistical grammars for open prompts as done in this section sound a little paradoxical. However, it turns out that, surprisingly to the most common intuition, statistical grammars seem to always outperform even very carefully designed rule-based grammars when enough training data is available. A respective study with four dialog systems and more than 2,000 recognition contexts was conducted in [120]. The apparent reason for this paradox is that in contrast to a general large-vocabulary language model trained on millions of word tokens, here, strongly context-dependent information was used, and statistical language models and classifiers were trained based only on data collected in the very context the models were later used in.

2.3.3 Robustness

Automatic speech recognition accuracy kept improving greatly over the last six decades since the first studies at Bell Laboratories in the early 1950s [27]. While some people claim that improvements have amounted to about 10% relative word error rate (WER[2]) reduction every year [44], this is factually not correct: It would mean that the error rate of an arbitrarily complex large-vocabulary continuous speech recognition task as of 2010 would be around 0.2% when starting at 100% in 1952. It is more reasonable to assume the yearly relative WER reduction being around 5% on average resulting in some 5% absolute WER as of today. This statement, however, is true for a trained, known speaker using a high-quality microphone in a room with echo cancellation [44]. When it comes to speaker-independent speech recognition in typical phone environments (including cell phones, speaker phones, Voice-over-IP, background noise, channel noise, echo, etc.) word error rates easily exceed 40% [145].

This sounds disastrous. How can a commercial (or any other) spoken dialog system ever be practically deployed when 40% of its recognition events fail? However, there are three important considerations that have to be taken into account to allow the use of speech recognition even in situations where the error rate can be very high [126]:

- First of all, the dialog manager does not use directly the word strings produced by the speech recognizer, but the product of the language understanding (SLU) component as shown in Fig. 1.1. The reader may expect that cascading ASR and SLU may increase the chance of failure since both of them are error-prone, and errors should grow rather than diminish. However, as a matter of fact, the combination of ASR and SLU has proven very effective when the SLU is robust enough to ignore insignificant recognition errors and still map the speech input to the right semantic interpretation.

  Here is an example. The caller says *I wanna speak to an associate*, and the recognizer hypothesizes *on the time associate* which amounts to 5 word errors

---

[2]Word error rate is a common performance metric in speech recognition. It is based on the Levenshtein (or edit) distance [64] and divides the minimum sum of word substitutions, deletions, and insertions to perform a word-by-word alignment of the recognized word string to a corresponding reference transcription by the number of tokens in said reference.

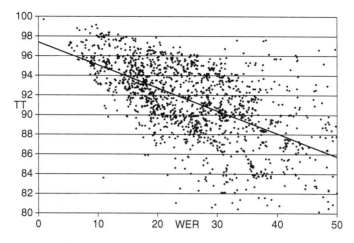

Fig. 2.1 Relationship between word error rate (abscissa) and semantic classification accuracy (True Total, ordinate)

altogether. Since the reference utterance has 6 words, the WER for this single case is 83%. However, the SLU component deployed in production was robust enough to interpret the sole presence of the word *associate* as an agent request and correctly classified the sentence as such resulting in no error at the output of the SLU module.

Figure 2.1 shows how, more globally, word error rate and semantic classification accuracy (True Total, see Sect. 3.2 for a definition of this metric) relate to each other. The displayed data points show the results of 1,721 experiments with data taken from 262 different recognition contexts in deployed spoken dialog systems involving a total of 2,998,254 test utterances collected in these contexts. Most experiments featured 1,000 or more test utterances to assure reliability of the measured values. As expected, the figure shows an obvious correlation between word error rate and True Total (Pearson's correlation coefficient is −0.61, i.e. the correlation is large [98]). Least-squares fitting a linear function to this dataset produces a line with the gradient −0.23 and an offset of 97.5% True Total that is also displayed in the figure. This confirms that the semantic classification is very robust to speech recognition errors reflecting only a fraction of the errors made on the word level of the recognition hypothesis.

Even though it may very well be due to the noisiness of the analyzed data, the fact that the constant offset of the regression line is not exactly 100% suggests that perfect speech recognition would result in a small percentage of classification errors. This suggestion is true since the classifier itself (statistical or rule-based), most often, is not perfect either. For instance, many semantic classifiers discard the order of words in the recognition hypothesis. This makes the example utterances

(1) Service interrupt

and

(2) Interrupt service

look identical to the semantic classifier while they actually convey different meanings:

(1) A notification that service is currently unavailable or a request to stop service

(2) A request to stop service

- It is well-understood that human speech recognition and understanding exploits three types of information: acoustic, syntactic, and semantic [45, 133]. Using the probabilistic framework typical for pattern recognition problems, one can express the search for the optimal meaning $\hat{M}$ (or class, if the meaning can be expressed by means of a finite number of classes) of an input acoustic utterance A in two stages:

$$\hat{W} = \arg\max_{W} p(W|A) = \arg\max_{W} p(A|W)p(W) \qquad (2.1)$$

formulates the determination of the optimal word sequence $\hat{W}$ given A by means of a search over all possible word sequences W inserted in the product of the *acoustic model* $p(A|W)$ and the *language model* $p(W)$. Similarly,

$$\hat{M} = \arg\max_{M} p(M|W) = \arg\max_{M} p(W|M)p(M) \qquad (2.2)$$

expresses the search for the optimal meaning $\hat{M}$ [36] based on the *lexicalization model* $p(W|M)$ and the *semantic prior model* $p(M)$ [78].

This two-stage approach has been shown to underperform a one-stage approach where no hard decision is drawn on the word sequence level [137]. In the latter case, a full trellis of word sequence hypotheses and their probabilities are considered and integrated with (2.2) [58, 84]. Despite its higher performance, the one-stage approach has not found its way into deployed spoken dialog systems yet because of primarily practical reasons, for instance:

- They are characterized by a significantly higher computational load (the search of an entire trellis requires extensively more computation cycles and memory than a single best hypothesis).
- Semantic parsers or classifiers may be built by different vendors than the speech recognizer, so, the trellis would have to be provided by means of a standardized API to make components compatible (see Sect. 2.6 for a discussion on standards of spoken dialog system component interfaces).

With reference to the different types of information used by human speech recognition and understanding discussed above, automatic recognition and understanding performance can be increased by providing as much knowledge as possible:

1. Acoustic models (representing the acoustic information type) of state-of-the-art speech recognizers are trained on thousands of hours of transcribed

speech data [37] in an attempt to cover as much of the acoustic variety as possible. In some situations, it can be beneficial to improve the effectiveness of the baseline acoustic models by adapting them to the specific application, population of callers, and context. Major phenomena which can require baseline model adaptation are the presence of foreign or regional accents, the use of the application in noisy environments as opposed to clean speech, and the signal variability resulting from different types of telephony connections, such as cell phone, VoIP, speaker phone, or landline.

2. In today's age of cloud-based speech recognizers [11], the size of language models (i.e. the syntactic information type) can have unprecedented dimensions: Some companies (Google, Microsoft, Vlingo, among others) use language models estimated on the entire content of the World Wide Web [18, 46], i.e., on trillions of word tokens, so, one could assume, there is no way to ever outperform these models. However, in many contexts, these models can be further improved by providing information characteristic to the respective context. For instance, in case of a directed dialog such as

> Which one can I help you with: Your bill, tech support, an order, an appointment, or a change to your service?

the a priori probabilities of the menu items (e.g. *tech support*) are much higher than those of terms outside the scope of the prompt (e.g. *I want to order hummus*). These priors have a direct impact on the optimality of the language model.

Even if only in-scope utterances are concerned, a thorough analysis of the context can have a beneficial effect on the model performance. An example: Many contexts of deployed spoken dialog systems are yes/no questions as

> I see you called recently about your bill. Is this what you are calling about today?

Most of the responses to yes/no questions in deployed systems are affirmative (voice user interface design best practices suggest to phrase questions in such a way that the majority of users would answer with a confirmation, as this has been found to increase the user confidence in the application's capability). As a consequence, a language model trained on yes/no contexts usually features a considerably higher a-priory probability for *yes* than for *no*. Thus, using a generic yes/no language model in contexts where *yes* is responded much less frequently than *no* can be disastrous as in the case where an initial prompt of a call routing application reads

> Are you calling about [name of a TV show]?

The likelihood of somebody calling the general hotline of a cable TV provider to get information on or order exactly this show is certainly not very high (even so, in the present example, the company decided to place this question upfront for business reasons), so, most callers will respond *no*. Using the generic yes/no language model (trained on more than 200,000 utterances, see Table 2.1) in this context turned out to be problematic since it tended to cause

Table 2.1 Performance of *yes* hypotheses in a yes/no context with overwhelming majority of *no* events comparing a generic with a context-specific language model

Language model	Training size	True Total of utterances hypothesized as *yes* (%)
Generic yes/no	214,168	27.3
Context-specific yes/no	1,542	77.4

substitutions between *yes* and *no* and false accepts of *yes* much more often than in regular yes/no contexts due to the wrong priors. In fact, almost three quarters of the cases where the system hypothesized that a caller responded with *yes* were actually recognition errors (27.3% True Total) emphasizing the importance of training language models with as much as possible context-specific information. It turned out that training the context-specific language model using less than 1% data than used for the generic yes/no language model resulted in a much higher performance (77.4% True Total).

- Last but not least, the amount and effect of speech recognition and understanding errors in deployed spoken dialog systems can be reduced by robust voice user interface design. There is a number of different strategies to this:

  - *rejection and confirmation threshold tuning*
    Both the speech recognition and spoken language understanding components of a spoken dialog system provide confidence scores along with their word or semantic hypotheses. They serve as a measure of likelihood that the provided hypothesis was actually correct. Even though confidence scores often do not directly relate to the actual *probability* of the response being correct, they relate to the latter in a more or less monotonous fashion, i.e., the higher the score, the more likely the response is correct. Figure 2.2 shows an example relationship between the confidence score and the True Total of a generic yes/no context measured on 214,710 utterances recorded and processed by a commercial speech recognizer and utterance classifier on a number of deployed spoken dialog systems. The figure also shows the distribution of observed confidence scores.

    The confidence score of a recognition and understanding hypothesis is often used to trigger one of the following system reactions:

    1. If the score is below a given *rejection threshold*, the system prompts callers to repeat (or rephrase) their response:

       I am sorry, I didn't get that. Are you calling from your cell phone right now? Please just say *yes* or *no*.

    2. If the score is between the rejection threshold and a given *confirmation threshold*, the system confirms the hypothesis with the caller:

       I understand you are calling about a billing issue. Is that right?

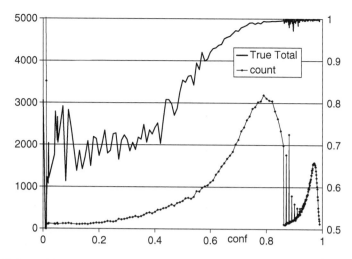

Fig. 2.2 Relationship between confidence score (abscissa) and semantic classification accuracy – True Total (ordinate, bold). The thin dotted line is the histogram of confidence values. The data is from a generic yes/no context

3. If the score is above the confirmation threshold, the hypothesis gets accepted, and the system continues to the next step.

Obviously, the use of thresholds does not *guarantee* that the input will be correct, but it increases the chance. To give an example, a typical menu for the collection of a cable box type is considered. The context's prompt reads

> Depending on the kind of cable box you have, please say either *Motorola*, *Pace*, or say *other brand*.

Figure 2.3 shows the relationship between confidence and True Total as well as the frequency distribution of the confidence values for this context. Assuming the following example settings[3]:

RejectThreshold = 0.07,
ConfirmThreshold = 0.85,

the frequency distribution of the box collection context can be used to estimate the ratio of utterances rejected, confirmed, and accepted.

In order to come up with an estimate for the accuracy of the box collection activity including confirmation (if applicable), re-confirmation, re-collection, and so on, one has to take into account that, in every recognition context, there are input utterances out of the system's action scope. In response to the question about the box type, people may say

---

[3] See Chap. 4 on how to determine optimal thresholds.

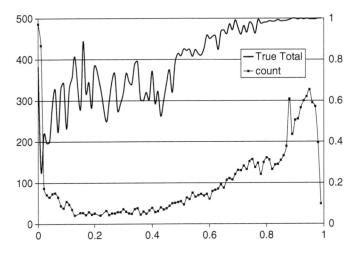

Fig. 2.3 Relationship between confidence score (abscissa) and semantic classification accuracy – True Total (ordinate, bold). The thin dotted line is the histogram of confidence values. The data is from a cable box collection context.

Table 2.2 Distribution of utterances among rejection, confirmation, and acceptance for a box collection and a yes/no context. The yes/no context is used for confirmation and, hence, does not feature an own confirmation context. Consequently, one cannot distinguish between TACC and TACA but only specify TAC. The same applies to TAW and FA

Event	Box collection (%)	Yes/No (confirmation) (%)
TACC	43.29	80.89
TACA	35.17	
TAWC	2.10	0.52
TAWA	0.03	
FAC	3.78	1.14
FAA	0.09	
FR	6.94	5.90
TR	8.61	11.56

I actually need a phone number, or the recognizer might have caught some side conversation or line noise, etc. Hence, when asking for how successful the determination of the caller's box type given the contexts' speech understanding performance is at the end, one will have to use the full set of spoken language understanding metrics discussed in Chap. 3 as demonstrated in Table 2.2.

In a standard collection activity that allows for confirmation, re-confirmation, re-collection, second confirmation, and second re-confirmation, there are 18 ways to correctly determine the sought-for information entity:

1. Correctly or falsely accepting[4] the entity without confirmation (TACA, FAA at collection),
2. Correctly or falsely accepting the entity with confirmation (TACC, FAC) followed by a correct or false accept of *yes* at the confirmation (TAC, FA).
3. Correctly or falsely accepting the entity with confirmation (TACC) followed by a true or false reject at the confirmation (TR, FR) followed by a correct or false accept of *yes* at the confirmation (TAC, FA).
4. ...

Instead of listing all 18 ways of determining the correct entity, the diagram in Fig. 2.4 displays all possible paths. Using the example performance measures listed in Table 2.2, one can estimate the proportional traffic going down each path and, finally, the amount ending up correctly (in the lower right box), see Fig. 2.5. Here, one sees the effectiveness of the collection/confirmation/re-collection strategy, since about 93% of the collections end up with the correct entity. The collection context itself featured a correct accept (with and without confirmation) of only 78.5%. This is an example for how robust interaction strategies can considerably improve spoken language understanding performance.

– *Robustness to specific input*
  In recognition contexts with open prompts such as the natural language call router discussed in Sect. 2.1, often, understanding models distinguishing hundreds of classes [115] are deployed. Depending on the very specifics of the caller response, the application performs different actions or routes to different departments or sub-applications. In an example, somebody calls about the bill. The response to the prompt

  > Briefly tell me what you are calling about today.

  could be, for example:

  > (1) My billing account number.
  > (2) How much is my bill?
  > (3) I'd like to cancel this bill.

---

[4]The author has witnessed several cases where a speech recognizer falsely accepted some noise or the like, and it turned out that the accepted entity was coincidentally correct. For example:

S: Depending on the kind of cable box you have, please say either *Motorola*, *Pace*, or say *other brand*.
C: <cough>
S: This was *Pace*, right?
C: That's correct.

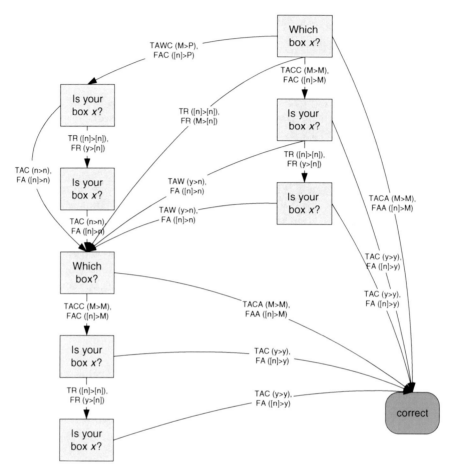

Fig. 2.4 Graph showing all successful paths of a disambiguation context with collection, re-collection, first and second confirmation. M=the correct box; P=a wrong box; [n]=noise or out-of-scope input; y=yes; n=no. $a > b$ represents an input event a that is understood as b by the speech recognition and understanding components

(4) Bill payment center locator.
(5) Change in billing.
(6) My bill is wrong.
(7) I wanna pay my bill.
(8) I need to change my billing address.
(9) Pay bill by credit card.
(10) Make arrangements on my bill.
(11) Seasonal billing.
(12) My bill.

All of these responses map to a different class and are treated differently by the application in how it follows up with the caller or routes the call to a destination.

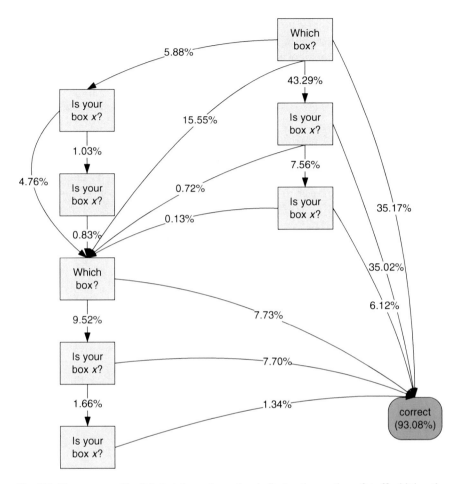

Fig. 2.5 The same as Fig. 2.4, but the path caption indicates the portion of traffic hitting the respective path

If, due to speech recognition and understanding problems, one of the specific responses (1–11) is classified as the generic one (12), this would be counted as an understanding error. The overall experience to the caller may, however, not be bad since the underlying high resolution of the context's classes is not known externally. An example conversation with this kind of wrong classification is

A1: Briefly tell me what you are calling about today.
C1: How much is my bill?
A2: You are calling about your bill, right?
C2: Yes.
A3: Sure. Just say *get my balance*, or *make a payment*. Or say, *I have a different billing question*.
C3: Get my balance.
A4: <presents balance>

(If there would not have been recognition problems, Turns A3, and C3 would have been bypassed). When looking at a number of example calls of the above scenario, there were 1,648 callers responding *yes* to the confirmation question A2 as opposed to 1,139 responding *no* (41%). This indicates that the disturbing effect of a substitution of a class by a broader class can be moderate. For the sake of completeness, when the classifier returned the right class, 11834 responses were *yes* and only 369 were *no* (3%).

– *Miscellaneous design approaches to improve robustness*
  There are several other voice user interface design techniques that have proven to be successful in gathering information entities such as [116]:

  • Giving examples at open prompts:

    Briefly tell me what you are calling about today.

    can be replaced by

    Briefly tell me what you are calling about today. For example, you can say *what's my balance?*

  • Offering directed back-up menu:

    Briefly tell me what you are calling about today.

    can be replaced by

    Briefly tell me what you are calling about today. Or you can say *what are my choices?*

  • Clear instructions of which caller input is allowed (recommendable in re-prompts):

    Have you already rebooted your computer today?

    can be replaced by

    Have you already rebooted your computer today? Please say *yes* or *no*.

  • Offer touchtone alternatives (recommendable in re-prompts):

    Please say *account information, transfers and funds,* or *credit or debit card information.*

    can be replaced by

    Please say *account information* or press 1, *transfers and funds* or press 2, or say *credit or debit card information* or press 3.

2.4   Dialog Management

After covering the system components speech recognition and understanding, Fig. 1.1 points at the *dialog manager* as the next block. In Sect. 1.1, it was pointed out that it "host[s] the system logic[,] communicat[es] with arbitrary types of backend services [and] generates a response ... corresponding to ... semantic symbols". This section is to briefly introduce the most common dialog management strategies, again with a focus on deployed solutions.

In most deployed dialog managers nowadays, the dialog strategy is encoded by means of a *call flow* that is a finite state automation [86]. The nodes of this automaton represent dialog activities, and the arcs are conditions. Activities can:

- Instruct the language generation component to play a certain prompt.
- Give instructions to synthesize a prompt using a text-to-speech synthesizer.
- Activate the speech recognition component with a specific language model.
- Query external backend knowledge repositories.
- Set or read variables,
- perform any type of computation, or
- Invoke another call flow as subroutine (that may invoke yet another call flow, and so on – this way, a call flow can consist of multiple hierarchical levels distributed among a large number of pages, several hundreds or even more).

Call flows are often built using WYSIWYG tools that allow the user to drag and drop shapes onto a canvas and connect them using dynamic connectors. An example sub-call flow is shown in Fig. 2.6.

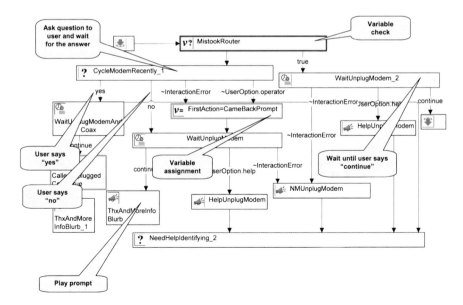

Fig. 2.6 Example of a call flow page

Call flow implementations incorporate features to handle designs getting more and more complex including:

- Inheritance of default activity behavior in an object-oriented programming language style (language models, semantic classifiers, settings, prompts, etc. need to be specified only once for activity types used over and over again; only the changing part gets overwritten; see Activities *WaitUnplugModem*, *WaitUnplugModem_2*, *WaitUnplugModemAndCoax* in Fig. 2.6 – they only differ in some of the prompt verbiage).
- Shortcuts, anchors, gotos, gosubs, loops.
- Standard activities and libraries collecting, for instance, phone numbers, addresses, times and dates, locations, credit card numbers, e-mail addresses, or performing authentication, backend database lookups or actions on the telephony layer.

Despite these features, complex applications are mostly bound to relatively simple human-machine communication strategies such as yes/no questions, directed dialog, and, to a very limited extent, open prompts. This is because of the complexity of the call flow graphs that, with more and more functionality imposed on the spoken language application, quickly become unwieldy. Some techniques to overcome the statics of the mentioned dialog strategies will be discussed in Chap. 4.

Apart from the call flow paradigm, there are a number of other dialog management strategies that have been used mostly in academic environments:

- Many dialog systems aim at gathering a certain set of information from the caller, a task comparable to that of filling a form. While one can build call flows to ask questions in a predefined order to sequentially fill the fields of the form, callers often provide more information than actually requested, thus, certain questions should be skipped. The *form-filling* (aka slot-filling) call management paradigm [89, 108] dynamically determines the best question to be asked next in order to gather all information items required in the form.
- Yet another dialog management paradigm is based on *inference* and applies formalisms from communication theory by implementing a set of logical principles on rational behavior, cooperation, and communication [63]. This paradigm was used in a number of academic implementations [8,33,103] and aims at optimizing the user experience by:

  - Avoiding redundancy.
  - Asking cooperative, suggestive, or corrective questions.
  - Modeling the states of system and caller (their attitudes, beliefs, intentions, etc.).

- Last but not least, there is an active community focusing on *statistical approaches* to dialog management based on techniques such as:

  - *Belief systems* [14, 139, 144]
    This approach models the caller's true actions and goals (that are hidden to the dialog manager because of the fact that speech recognition and understanding

are not perfect). It establishes and updates an estimate of the probability distribution over the space of possible actions and goals and uses all possible hints and input channels to determine the truth.

- *Markov decision processes/reinforcement learning* [56, 66]
  In this framework, a dialog system is defined by a finite set of dialog states, system actions, and a system strategy mapping states to actions allowing for a mathematical description in the form of a Markov decision process (MDP). The MDP allows for automatic learning and adaptation by altering local parameters in order to maximize a global reward. In order to do so, an MDP system needs to process a considerable number of live calls, hence, it has to be deployed, which, however, is very risky since the initial strategy may be less than sub-optimal. This is why, very often, simulated users [7] come into play, i.e. a set of rules representing a human caller that interacts with the dialog system initializing local parameters to some more or less reasonable values. Simulated users can also be based on a set of dialog logs from a different, fairly similar spoken dialog system [48].

- *Partially observable Markov decision processes* [143]
  While MDPs are a sound statistical framework for dialog strategy optimization, they assume that the dialog states are *observable*. This is not exactly true since caller state and dialog history are not known for sure. As discussed in Sect. 2.3.3, speech recognition and understanding errors can lead to considerable uncertainty on what the real user input was. To account for this uncertainty, partially observable Markov decision processes (POMDPs) combine MDPs and belief systems by estimating a probability distribution over all possible caller objectives after every interaction turn. POMDPs are among the most popular statistical dialog management frameworks these days. Despite the good number of publications on this topic, very few deployed systems incorporate POMDPs. Worth mentioning are those three systems that were deployed to the Pittsburgh bus information hotline in the summer of 2010 in the scope of the first Spoken Dialog Challenge [13]:

  - AT&T's belief system [140].
  - Cambridge University's POMDP system [130].
  - Carnegie Mellon University's benchmark system [95] based on the Agenda architecture, a hierarchical version of the form-filling paradigm [102].

2.5   Language and Speech Generation

(Natural) language generation [26] refers to the production of readable utterances given semantic concepts provided by the dialog manager. For example, a semantic concept could read

```
CONFIRM: Modem=RCA
```

i.e., the dialog manager wants the speech generator to confirm that the caller's modem is of the brand RCA. A suitable utterance for doing this could be

> You have an RCA modem, right?

Since the generated text has to be conveyed over the audio channel, the *speech generation* component (aka speech synthesizer, text-to-speech synthesizer) transforms the text into audible speech [114].

Language and speech generation as described above are typical components of academic spoken dialog systems [94]. Without going into detail on the technological approaches used in such systems, it is apparent that both of these components come along with a certain degree of trickiness. Since language generation has to deal with every possible conceptual input provided by the dialog manager it is either based on a set of static rules or relies on statistical methods [39, 60]. Both approaches can hardly be exhaustively tested and lack predictability in exceptional situations. Moreover, the exact wording, pausing, or prosody can play an important role for the success of a deployed application (see examples in [116]). Rule-based or statistical language generation can hardly deliver the same conversational intuition like a human speaker. The same criticism applies to the speech synthesis component. Even though significant quality improvements have been achieved over the past years [57], speech synthesis generally lacks numerous subtleties of human speech production. Examples include:

- Proper stress on important words and phrases:

  > S: In order to check your connection, we will be using the *ping* service.

- Affectivity such as when apologizing:

  > S: Tell me what you are calling about today.
  > C: My Internet is out.
  > S: I am sorry you are experiencing problems with your Internet connection. I will help you getting it up and running again.

- Conveying cheerfulness:

  > S: Is there anything else I can help you with?
  > C: No, thank you.
  > S: Well, thank *you* for working with me!

Even though there is a strong trend towards affective speech processing evolving over the last 5 years potentially improving these issues [85], the general problem of speech quality associated with text-to-speech synthesis persists. Highly tuned algorithms trained on large amounts of high-quality data with context awareness still produce audible artifacts, not to speak of certain commercial speech synthesizers that occasionally produce speech not even intelligible.

All the above arguments are the reasons why deployed spoken dialog systems hardly ever use language and speech generation technology. Instead, the role of the voice user interface designer comprises the *writing and recording of prompts*.

That is, every single system response is carefully worded and then recorded by a professional voice talent in a sound studio environment. At run-time, the spoken dialog system simply plays the pre-recorded prompt producing optimal sound quality[5]. Dynamic contents (such as the embedding of numbers, locations, e-mail addresses, etc.) can be implemented in a concatenative manner with pre-recorded contents as well. Only in instances where the nature of the presented contents is unpredictable or of a prohibitive complexity (such as with last names in a phone directory application on a large and frequently changing set of destinations), speech synthesis has no alternative.

In spite of the clear advantage of the prerecorded prompt approach, it features the clear disadvantage that every single prompt needs to be formulated and recorded covering every possible situation that can arise in the course of every dialog activity including, e.g.:

- The announcement prompt (the introductory part of the activity).
- Re-announcement prompts.
- Announcement-interrupted prompt(when the caller interrupts the announcement).
- Question prompt.
- Hold prompt (a caller asks the system to hold on).
- No-input, no-match, etc. prompts for the hold role.
- Hold-return prompt (resumes the interaction after a hold).
- No-input prompts (when the caller does not say anything).
- No-match prompts (when the caller caused a reject).
- Confirmation prompts (when the speech input needs to be confirmed).
- No-input, no-match, etc. prompts for the confirmation role.
- N-best prompts (when more than one recognition hypothesis is used for the confirmation).
- Help prompt (when the caller asked for more information).
- Operator prompt (when the caller asked for an agent).
- Expert prompt (when the caller is an expert user).
- Repeat prompt (when the caller asked to repeat the information), or
- Technical-difficulty prompt.

Consequently, deployed systems of regular complexity usually require thousands, sometimes tens of thousands of pre-recorded prompts. For example, the Internet troubleshooting application described in [6] currently comprises 10,573 prompts with a total duration of 33 h. As a result, the professional recording of prompts plays a major role for the overall cost and time of building an application. Presumably trivial projects such as switching the voice talent or localizing an existing spoken dialog system to another language [118] can become prohibitive.

---

[5]This approach occasionally tricks callers in that they assume to be talking to a live person.

2.6   Voice Browsing

It became obvious to the speech industry that there was a need for standardized speech interfaces for spoken dialog systems only after the market saw an uptake in the number of speech applications that were introduced into the market, accompanied by the burgeoning number of speech vendors and consumers of such commercial spoken dialog systems. Given that speech recognizers, text-to-speech systems, telephony infrastructure, dialog managers, backend infrastructure, and the actual applications are potentially built by different companies in the first place, by standardizing how these components talk to each other, architecting and building solutions became much easier.

A great step towards the modularization of spoken dialog system components was the introduction of a proxy component, the *voice browser* [61]. It takes over the communication layer between speech recognition and synthesis on the one hand and language understanding and generation on the other as shown in Fig. 2.7. In an alternative architecture, speech recognition and understanding are coupled, so the voice browser communicates directly with the dialog manager (see Fig. 2.8).

As its name suggests, the voice browser plays a role similar to a web browser which most often communicates with a human client on the one hand and a web server on the other. In this analogy (see Fig. 2.9), speech recognition (and, potentially, understanding) functions as input device of the voice browser which, in the web world, are keyboard, mouse, camera, and other input channels communicating with the web browser. Output device of the voice browser is the speech synthesizer that replaces the screen, loudspeakers, and other output channels used by a web browser. On the internal side, a voice browser communicates with the dialog manager (or with the spoken language understanding and generation components that are directly controlled by the dialog manager) playing the role of

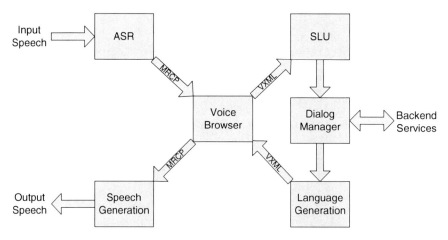

Fig. 2.7 General diagram of a spoken dialog system with voice browser

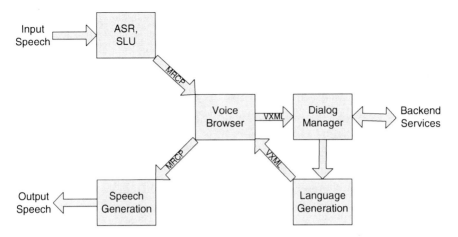

Fig. 2.8 General diagram of a spoken dialog system with voice browser; ASR and SLU coupled

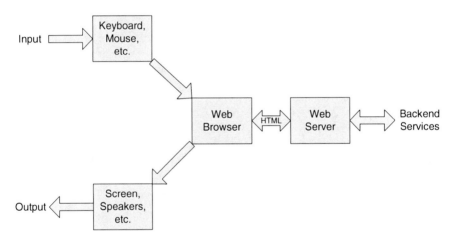

Fig. 2.9 General diagram of a web browser

the web server in the web-based world. In fact, modern implementations of dialog managers are web applications making use of standard web servers such as Apache or Internet Information Services as well as common programming environments as Java Servlets, PHP, or .NET. Very much like their web counterparts, the components of Figs. 2.7 and 2.8 can be distributed over local and wide area networks communicating via HTTP and other standard protocols (in fact, the applications the author was working on in the past years – see e.g. [118, 120, 121, 123, 124] for details – were hosted on infrastructure in New York, New Jersey, Pennsylvania, California, Georgia, among others).

Inspired by the strength of standardization in the web world where the Hypertext Markup Language (HTML) serves as primary markup language for web pages, and almost all available browsers and web content generators adhere to this standard, in 1999, a forum based on a selection of the most advanced speech research laboratories (AT&T, IBM, Lucent, and Motorola) was founded to develop a markup language for spoken dialog systems [109]. Based on the general definition of the Extensible Markup Language (XML), the new standard was branded VoiceXML, and soon after releasing Version 1 in 2000, control was handed over to the World Wide Web consortium (W3C) that made VoiceXML a W3C Recommendation in 2004 [72].

VoiceXML specifies, among other features:

- Which prompts to play (TTS or pre-recorded audio files).
- Which language or classification models (aka grammars, see Sect. 2.3) to activate (speech and touch tone).
- How to record spoken input or full-duplex telephone conversations.
- Control of the call flow.
- Telephony features for call transfer or disconnect.

VoiceXML was meant to open the entire feature space of the World Wide Web to the domain of spoken dialog systems. In this way, it was to:

- Minimize the number of transactions between voice browser and dialog manager (see Sect. 2.7 on how crucial and demanding real-time ability can be in distributed spoken dialog systems) – simple dialog systems can be implemented as a single VoiceXML page.
- Separate application code (VoiceXML) from low-level platform code (that can be in whatever programming language, or come along as a precompiled application).
- Allow for portability across different VoiceXML-compliant platforms (for both voice browsers and dialog managers).
- VoiceXML can be static (like static HTML), or dynamic (produced by dynamic web content generators such as PHP, CGI, Servlets, JSP, or ASP.NET).

Certainly, the most important step towards the modularization of spoken dialog systems was the specification of VoiceXML as the interface between dialog manager and voice browser. However, the internals of the voice browser itself, which had been originally introduced to serve as a proxy for proper communication between dialog manager and speech recognition and generation, still required well-defined interfaces. Again, this was because vendors of browser, ASR, and TTS in a single bundle could be multilateral, and there was a high demand for standardization to make components compatible with each other [19]. The response to this demand was the Media Resource Control Protocol (MRCP) published in 2006 by the Internet Society as an RFC (Request for Comments) [106]. MRCP controls media resources like speech recognizers and synthesizers and uses streaming protocols such as the Session Initiation Protocol (SIP), widely deployed in Voice-over-Internet-Protocol telephony [51].

2.7   Deployed Spoken Dialog Systems are Real-Time Systems

The heavy use of distributed architecture (see Fig. 3.2 for a high-level diagram of a deployed spoken dialog system's architecture including infrastructure to measure performance) requires a lot of attention to the real-time ability of the involved network machinery. In order to understand what *real-time processing* means in the context of deployed spoken dialog systems, one can use human-to-human phone conversations as a standard of comparison.

The average pause length between interaction turns is about 250 ms [15, 42], and the average tolerance interval, i.e., the time after which the conversational partner feels obliged to speak, is approximately 1 s for American English speakers [50]. This means that the time lag between the moment when the caller stops and that when the system starts speaking should not be considerably longer than 1 s. If this requirement is not fulfilled, callers tend to repeat themselves assuming the system missed their response to a prompt (Turn 1). This repetition, however, may fall into the time scope of the next interaction turn (Turn 2) and, hence, may be interpreted as the response to the question of Turn 2. It is possible that the caller only heard snippets (or possibly nothing at all) of Turn 2's prompt, since, often, question prompts allow for so-called barge-in: Callers can respond at any time during the prompt and do not have to wait until the end of a possibly lengthy prompt allowing expert users to quickly navigate through a speech menu.

Table 2.3 displays an example conversation taken from a call routing application. The application was tuned to minimize handling time (around 37 s on average) producing substantial cost savings considering a volume of about 4 million calls per month.

This conversation features major glitches mainly because of the system taking too long to respond:

- The caller utters a response (3), waits for 1.3 s (4) to decide that the system either did not hear or is still listening, and qualifies her former response by saying *Technical* (5, 6). At this moment, the speech recognizer has already stopped listening, and the dialog manager is preparing the next context. In fact, the first 200 ms of the caller response (*Tech*) still fall into Context 1. The remaining part of the utterance (*nical*) coincides with the next context's system prompt that does not get played at all for being interrupted by the caller, and the corrupted utterance is interpreted in the scope of Context 6. The system receives a response that is out-of-scope for Context 6 (the fragment *nical* cannot be interpreted) and, consequently, re-prompts (8) by saying *I didn't get that...*
- The caller assumes the system is still in Context 1 and did not understand her response, so, she repeats her former input (9), pauses again for 1.2 s (10) without any system response and qualifies her answer by saying *Tech support* (11). The latter, however, again coincides with a system response to Input 9 (*Phone, sure*) and gets ignored since the system is not listening during this indirect confirmation prompt.

Table 2.3 Example conversation in a call router application showing problems arising due to latency. Gray parts of the system prompt are not played due to barge-in by the caller

ID	Time/s	System	Caller
1	0	Briefly tell me what you are calling about today. For example: *I want to order new services.*	
2	4.7	<2.5 s silence>	
3	7.2		Telephone.
4	8.0	<1.3 s silence>	
5	9.3		Tech...
6	9.5	Which one can I help you with: your bill, tech support, an order, an appointment, or a change to your service?	...nical.
7	10	<1.9 s silence>	
8	11.9	I didn't get that. Just say *my bill* or press 1, *tech support* or press 2, *an order* or press 3, *an appointment* or press 4. Or say *make a change to my service* or press 5.	
9	18.1		Telephone.
10	18.9	<1.2 s silence>	
11	20.1	Phone, sure.	Tech support.
12	21.4	<0.8 s silence>	
13	22.2	Just say *my bill* or press 1, *tech support* or press 2, *an order* or press 3, *an appointment* or press 4. Or say *make a change to my service* or press 5.	
14	31.8		Tech support.
15	32.7	<0.8 s silence>	
16	33.5		Tech sup...
17	34.0	Are you having trouble with the dial tone on your phone?	...port.
18	34.4	<3.5 s silence>	
19	37.9	I didn't get that. If you're having trouble with the dial tone say *yes*, otherwise, say *no.*	
20	40.5		Tech support.
			Tech support.
21	43.8	<1.9 s silence>	
22	45.7	OK. Let me get someone on the line to help you.	
23	48.0	<1.0 s silence>	
24	49.0		Thank you.

- After another silence to load the next prompt (12), the system starts speaking (13) offering menu options including the one just ignored (Tech support). The patient caller repeats herself (14), waits for 0.8 s (15) and repeats herself once again (16, 17). In the meantime, the system has already interpreted Response 14 and moves on to the next context while the speaker already started speaking (16). Again, the prompt gets interrupted right away, and the recognizer only captures the second part of the response (*port*) which cannot be successfully interpreted.
- Consequently, the system apologizes and replays the question (19). The caller assumes she is still in Context 13, and, therefore, interrupts the prompt repeating her former response twice (20). Since her input still does not answer the question, the system gives up according to the application's policy and escalates to a human operator (22).

The reader may want to argue that the speech understanding problems could have been reduced by:

1. Overcoming technical hurdles making the system listen without even slight interruptions (thereby avoiding the cut user inputs 5/6 and 16/17).
2. Revisiting the barge-in behavior of certain prompts (e.g. forcing the caller to listen to the first seconds of 6 and 17).

(1) is in the responsibility of the technology vendors (i.e. the developers of speech recognizer and voice browser) which, as discussed above, are usually companies different from the ones building the applications, making it a hard problem to tackle. (2) is in the court of the voice user interface designers, but there are also a number of drawbacks to forcing callers to listen to extended prompts, inter alia, an increase of average handling time and the fact that speech input may not be acknowledged at all (exemplified by Turn 11 in Table 2.3), in turn resulting in potential understanding problems.

Generally, a significant reduction of latency most probably would have saved the above sample conversation to begin with. To understand what it takes to make deployed spoken dialog systems in a distributed environment real-time-able, one needs to look at all the actions performed between the moment when a caller's speech is over and when the system response starts playing (considering the architecture shown in Fig. 2.8):

As shown in Table 2.4, there are three types of contributors to the overall latency, constant (C), server-load-dependent (S), and network-dependent (N) ones. The single constant contributor, the complete recognition time-out (i.e. the duration the recognizer waits after the caller stops speaking until deciding that the utterance is over), cannot be altered without compromising recognition and understanding accuracy due to false end-point detection (in fact, there is extensive scientific work dedicated to the determination when to take turn based on various clues such as prosody, syntax, semantics, or pragmatics [53, 82, 131]). Latency caused by server overload can be reduced by carefully balancing load among available servers or by upgrading the stock of available computational resources connecting additional machines. Finally, the network needs to be laid out to accommodate guaranteed response times of a magnitude lower than 100 ms round-trip delay (consider that a single voice browser/dialog manager turn can involve up to seven network transactions or even more depending on the specific communication protocol). This response time may not exceed a certain maximum threshold (e.g., 100 ms) even in case of occasional high-load situations.

To get a rough idea of the required network capacity in such a real-time system, the example scenario referred to in Fig. 1.4 is considered where:

- In peak situations, a customer service hotline receives some $n = 20,000$ calls per hour.
- Every single of these calls is processed by the call routing application mentioned earlier in this chapter.

Table 2.4 Steps performed by a deployed spoken dialog system between a caller stops talking and the system starts responding. C is a constant contribution to latency, while S and N are variable durations depending on server load and network speed, respectively

step	$C|S|N$
Complete recognizer time-out (this is the time the recognizer waits until deciding that the speaker utterance is over and that the silence is not a natural speaking pause) (ASR)	C (1,000 ms)
Completing speech recognition and delivering the recognition hypothesis (ASR)	S
Classifying the recognition hypothesis and delivering the semantic hypothesis (SLU)	S
Returning recognition and semantic hypotheses over the network to the voice browser	N (<5 ms LAN; <100 ms WAN)
The voice browser decides whether to ignore the recognition event based on the semantic hypothesis (in so-called *hot-word* contexts, the application is to ignore all user inputs but a number of predefined classes in order not to interrupt the conversation unnecessarily – see e.g. Context 11 in Table 2.3)	S
In regular contexts, the voice browser forwards recognition and semantic hypotheses over the network to the dialog manager	N
The dialog manager processes the voice browser's output, navigating the call flow, accessing backend services if required, and preparing the system's response (language generation)	S (3 s with, 100 ms without backend)
The dialog manager sends the next request to the voice browser over the network providing information about what prompt to play, which speech recognition and understanding models to load, and a number of additional parameters such as time-outs, sensitivity, confidence thresholds, etc. (for details about these, see Sect. 2.3)	N
The dialog manager request gets compiled (or interpreted) by the voice browser	S
All required prompts (audio files) are requested over the network (they are usually located on a separate media server). Alternatively, the prompt text is sent over the network to a text-to-speech module	N[a]
If applicable, the text-to-speech module generates an audio signal (speech generation)	S
The audio signal or file is sent back to the voice browser (or directly to the prompt player) over the network	N[a]
Speech recognition and understanding models are requested over the network (they are usually located on a separate media server)	N[a]
Speech recognition and understanding models are sent back to the voice browser (or directly to the speech recognizer) over the network	N[a]
ASR and SLU modules are compiled by providing speech recognition and understanding models	S[a]
ASR starts listening	–
The prompt starts playing	–

[a]Indicates that this contribution does not apply when server file caching is active

- One call requires 19.1 transactions between voice browser and dialog manager on average (measured on data from July 2010).
- A single transaction averages at 3,463 bytes sent from the dialog manager and 700 bytes the other way (measured on data from July 2010).

Table 2.5 Network throughput produced by a number of applications hosted on two data centers (one for the voice browsers, one for the dialog managers) connected by a single wide area network connection

Application	Customer	Throughput/(Mbit/s)
Call router	A	2.81
Internet troubleshooting	A	1.80
Cable TV troubleshooting	A	0.80
Digital phone troubleshooting and FAQ	A	0.03
FAQ (about settings and new cable equipment)	A	0.07
Customer survey after speaking to a human agent	A	0.78
Call-back application after outage clearance	A	0.02
Internet troubleshooting	B	0.21
Cable TV troubleshooting	B	0.33
Sum		6.84

Using these values, one can compute the average load for the dialog manager outbound connection as

$$L = 20000 \cdot 19.1 \cdot 3463 \text{ bytes/hour} = 2.81 \text{ Mbit/s}. \qquad (2.3)$$

While this amount sounds non-critical assuming that reliable high-speed Internet connections are available for at least 10 Mbit/s, one has to consider that there may be other applications sharing the same network connection. Specifically, as the example application is a call router, it routes callers to human operators or other spoken dialog systems. When these other systems' voice browsers and dialog managers are hosted in the same facilities as those of the call router, most often, they will share the network connection. In the case of the present example, Table 2.5 shows which applications were sharing the network connection with the call router and which expected throughput each of them produced.

Moreover, transactions are not evenly distributed during the 1-h time frame. Similar to what was discussed in Sect. 1.2, one can calculate the likelihood that transactions overlap in time, and, based on that, what the expected network latency caused by overlapping transactions would be.

Chapter 3
Measuring Performance of Spoken Dialog Systems

Abstract Key to the evaluation of spoken dialog systems and the prerequisite to tuning these systems is to properly measure their performance. This chapter reviews common performance metrics distinguishing between subjective, objective, observable, and hidden domains. A special focus is placed on spoken language understanding performance metrics and on the architecture required to gather the data necessary to calculate these metrics.

Keywords Evaluation infrastructure • Hidden metrics • Objective metrics • Observable metrics • Performance metrics • Semantic annotation • Speech performance analysis metrics • Subjective metrics

The previous chapter reviewed multiple techniques for building spoken dialog systems with a focus on the challenges of deploying such systems to the real world. If a fair comparison between these (and possibly other) techniques is to be drawn, their performance needs to be measured in some way. The present chapter is to review state-of-the-art methods to apply performance measures to spoken dialog systems briefly commenting on challenges and current trends in this endeavor.

3.1 Observable vs. Hidden

Without doubt, the main objective of deployed spoken dialog systems is to offer callers self-service options that, under normal circumstances, a human agent would have offered [24]. That is, spoken dialog systems attempt to automate a human's task. Accordingly, these systems' performance should be tied to some measure of the effectiveness of this attempt, the most direct of which is the automation rate (aka completion rate, deflection rate, or Tier 1 performance).

While the concept of an automation rate, i.e. the average number of calls that successfully completed the interaction with the caller divided by the total number of calls, sounds like a straightforward measure, it is actually not. Looking at two example applications:

1. A call router that is intended to route a call to that department best matching the call reason.
2. a technical support application for cable TV troubleshooting,

how does one tell whether a call was automated?

1. According to the objective of a call router, an automated call would be one that ended up at the right destination, i.e., the right department, agent, or automated application. It is indeed possible to reliably say whether or not a call ended up at *a* department, agent, or automated application. However, how does one tell whether or not this destination was the *right* one? Common practice is to assume that *every* routed call is a correctly routed one. So, when the dialog manager believes to have captured the call reason and routes the call, this would be considered automated. Only in the (rare) case that the application is not able to determine the call reason (due to repeated recognition problems, the caller asking for human assistance, the caller not making any input or hanging up), the call would be classified as *not automated*. The fraction of the latter is potentially very small. Effective call routers can have as little as 5% or 10% non-automated calls which sounds great when compared to other spoken dialog systems (see below).

   However, when one looks at what happens after callers were "successfully" routed to their destination, it turns out that there may be a considerable number of calls whose routing destination does not match their needs leading to cross-routes among the different departments inside the call center network. Sometimes, the percentage of callers experiencing a cross-route exceeds 10% which makes the real total number of non-automated calls be in the 20% range.

   Here, a typical problem becomes obvious: To evaluate system performance, one often has to rely on facts that are directly measurable by the system (such as whether and where calls were routed, how long the call was, how many callers were cross-routed). These facts are referred to as *observable* facts [125]. On the other hand, there are facts the system does not know (what is the caller's goal, what did the caller say, which speech inputs were not heard, etc.), referred to as *hidden* facts.

   Often, hidden facts are mistaken as observable ones, when, for instance, people report on the number of agent requests by looking at log data. What can be seen in log data, however, are those events that were *interpreted* as agent requests, a set that differs (potentially significantly) from the set of actual agent requests. This is due to a number of speech recognition and understanding mistakes that can have led to events falsely interpreted as agent requests, or, contrarily, actual agent requests misinterpreted or ignored. In Sect. 3.2, it will be shown how these types of events can be measured by means of speech performance analysis metrics.

2. Technical troubleshooting applications aim at resolving a problem the caller experiences. A resolution can be obtained by executing certain actions that directly impact the network-connected equipment suffering the problem using some sort of backend integration (ping service, refresh or reboot of equipment, deploy new firmware). Alternatively, the caller can be instructed to manually carry out certain activities supposed to resolve the issue. In both cases, it often takes several minutes until the problem disappears, a time in which the system usually remains in a wait state, i.e., it expects a caller input to continue. A typical interaction is, e.g.:

> S: We're almost done! Now, go ahead and try to connect to the web the way you usually do by opening a web browser like Internet Explorer or another program. When you do connect and are able to see your homepage, say *continue*.
> C: Continue.
> S: To confirm, you *do* see your home page now, right?
> C: Yes.
> S: Now, to make sure you have a good connection, I need you to go to a different web site. So, towards the top of your browser is an address bar where you can type in a web address. Go ahead and click there. The address is `www.speechcycle.com`. That's WWW dot SpeechCycle dot com. Do you need to hear that again?
> C: No.
> S: After typing in that address, please tell me one of these three things: *I am connected, still no Internet*, or *repeat the address*.
> C: I am connected.
> S: Just to confirm, you *are* connected, is that right?
> C: Yes.

The major part of this conversation is to confirm that the caller's problem has been resolved and the call can be flagged as automated. However, many callers, aware of that they are talking to a machine, are not patient enough to complete the dialog until its very end but hang up once they have discovered that their problem has disappeared. In this case, there is no way to tell whether callers hung up out of frustration, impatience, or because of the problem being resolved. Unfortunately, in some applications, a significant portion of the calls (20% or more) end with the caller hanging up in situations where it is unclear whether the problem was resolved or not. This is yet another example for the fuzziness of the notion automation rate.

In addition to the automation rate, there is a number of common objective metrics used to evaluate the performance of spoken dialog systems, e.g.:

- Average handling time [127].
- Number of operator requests (hidden – extrapolated by observable events) [73].
- Number of hang-ups [17].
- "speech errors" (number of rejects, disconfirmations, time-outs, etc. some of which are hidden but get extrapolated by observable events) [101].
- Exit analysis (which category or state was the call in when it finished?) [16].
- Cost savings (specially important in commercial applications) [123].

The specifics of these metrics are not in the scope of this work. However, for the discussions in Chap. 4, it is crucial to agree on a scalar observable metric (that can very well be some combination of the above and other metrics) to be able to adapt and optimize a deployed spoken dialog system. For the tuning of the speech recognition and understanding components of the system, one also needs to consult hidden speech performance metrics discussed in Sect. 3.2 in further detail.

3.2   Speech Performance Analysis Metrics

Throughout the previous chapters and sections of this work, the notion of *speech (recognition and understanding) accuracy* or *performance* or *errors* has been repeatedly used without further detail on how they are defined. Since speech recognition and understanding (together with the dialog manager) play the most important roles concerning the functionality of a spoken dialog system, the proper description of their performance is crucial. Without going into deep detail ([125] contains thorough motivation and discussion of this topic), the most important definitions are reiterated at this point.

In Sect. 2.3.3, it was explained why errors a speech recognizer produces at the word level do not necessarily propagate to the dialog manager due to the error robustness of the spoken language understanding component, and therefore are only a rather weak measure to describe the performance of a dialog system's input channel. Consequently, the measuring of speech recognition and understanding performance is defined in the *semantic* domain. According to the general architecture of spoken language understanding, the semantic representation of a spoken input can be a complex hierarchy (see e.g. the review in [137]). However, deployed systems almost exclusively use a flat topology, i.e., the semantic representation of a caller's utterance is one out of a (possibly infinite) set of classes. This topology covers, among others, the following common dialog paradigms:

- Yes/no questions
- Menus
- Open prompts (How-May-I-Help-You style)
- Date, amount, location, phone number, credit card information, e-mail addresses, etc.
- User initiative.

Even though this topology was called *flat* referring to the fact that a single class is used to describe the semantic content of a given input utterance, the underlying semantic representation can adhere to a complex hierarchy as exemplified by the screenshot in Fig. 3.1. This figure displays a software used to *annotate*[1] a set of

---

[1]Annotation refers to the (mostly manual) process of assigning a semantic class to a given transcription, i.e. textual representation of an utterance.

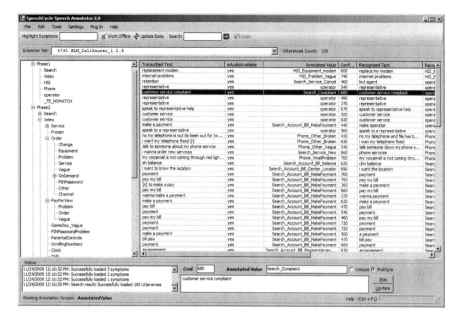

Fig. 3.1 Example of a semantic annotation software

input utterances (rows in the table on the right). The set of classes is shown on the left in form of a tree whose leaves in conjunction with all branches necessary to reach the leaf form the semantic class. Using "_" as (arbitrary) delimiter between branches, one example of the displayed classes is

    Phase2_Video_Order_Equipment

One of the problems with representing a hierarchical semantic structure as a set of flat semantic classes is that every type of error is counted the same, independently of how invasive it would be. The probably non-essential substitution

    Phase2_Video_Order_Other ⟹ Phase2_Video_Order_Vague

is counted the same as

    Phase1_operator ⟹ Phase2_Video_ParentalControls

(see Sect. 2.3.3 for an example on how harmless certain substitutions are).

Furthermore, the flat topology does not directly cover situations where multiple pieces of information are collected from a single user utterance (*I want to pay my bill and change my home address*). However, since this type of multiple inputs is very rare, they are usually covered by a single "multiple" class of the most specific common branch (in the last example, it is Phase2_Search_AccountBill_Multiple).

As introduced in [117], a set of utterances for which the semantic annotations as well as the classes returned by the spoken language understanding component (in a production deployment or an experimental lab environment) are known can be split according to the following criteria:

1. *Scope.* An utterance is covered by one of the canonical classes defined in the class set of the respective recognition context (in scope) or not (out of scope). Out-of-scope utterances include noise and any type of utterances that are not handled by the dialog system logic of the recognition context in question.
2. *Acceptance.* The spoken language understanding component can either deem the utterance in-scope and, accordingly, *accept* it or, contrarily, *reject* it. As already discussed in Sect. 2.3.3, a low recognition/understanding confidence score can also suggest to reject the utterance since the recognition hypothesis is most likely wrong.
3. *Correctness.* When an in-scope utterance was accepted, this criterion is to determine whether the predicted class was identical to the annotated one (correct) or not (false).
4. *Confirmation.* This determines whether an event was confirmed.

With growing complexity of the interaction, performance metrics can be introduced to cover typical events whose frequency of occurrence is to be measured. The least complex interaction is one that features a single in-scope class, i.e., it is a binary classification task. Examples for this scenario are announcement contexts where callers are not supposed to say anything with the only exception of an agent request that some business policies require to be active at all time throughout an application. Here, it is sufficient to know the scope and the acceptance of an utterance to describe all possible events:

- When an in-scope utterance gets accepted it is called a True Accept (TA).
- When an in-scope utterance gets rejected it is called a False Reject (FR).
- When an out-of-scope utterance gets accepted it is called a False Accept (FA).
- When an out-of-scope utterance gets rejected it is called a True Reject (TR).

Table 3.1 shows a more comprehensible diagram of these binary classification metrics. An overview about all performance metric acronyms used in this work is given in Table 3.2.

Most recognition contexts are, indeed, not of binary nature, and, hence, the fact whether an in-scope utterance was accepted does not suffice to express whether the predicted class matched the actual (annotated) class. This is why one distinguishes between *correct* accepts and *wrong* accepts (aka substitutions). According to the naming convention, these cases are called True Accept Correct (TAC) and True Accept Wrong (TAW), respectively. An illustration is given in Table 3.3.

Table 3.1 Spoken language understanding performance metrics – the case of binary classification

	A	R
I	TA	FR
O	FA	TR

Table 3.2 Spoken language understanding performance metrics – acronyms

I	In-Grammar
O	Out-of-Grammar
A	Accept
R	Reject
C	Correct
W	Wrong
Y	Confirm
N	Not-Confirm
TA	True Accept
FA	False Accept
TR	True Reject
FR	False Reject
TAC	True Accept Correct
TAW	True Accept Wrong
FAC	False Accept Confirm
FAA	False Accept Accept
TACC	True Accept Correct Confirm
TACA	True Accept Correct Accept
TAWC	True Accept Wrong Confirm
TAWA	True Accept Wrong Accept
TT	True Total
TCT	True Confirm Total

Table 3.3 Spoken language understanding performance metrics – the case of non-binary classification

	A		R
	C	W	
I	TAC	TAW	FR
O	FA		TR

The table cells highlighted in gray are the "good" metrics – that is, whenever an utterance is in scope, it should be correctly accepted (TAC), otherwise it should be rejected (TR). To describe the spoken language understanding performance of a recognition context in general, one therefore combines the good metrics to the overall metric True Total defined as

$$TT = TAC + TR. \tag{3.1}$$

Finally, there are recognition contexts with enabled confirmation (as introduced in Sect. 2.3.3). Here, it is worthwhile to quantify how effective the detection of utterances to be confirmed is as compared to the other types (accepts and rejects). Accordingly, one splits all sets of accepted utterances (TAC, TAW, FA) into confirmed and non-confirmed (directly accepted) resulting in the six additional metrics TACC, TACA, TAWC, TAWA, FAC, FAA shown in Table 3.4 (for their expansion, see Table 3.2).

Table 3.4 Spoken language
understanding performance
metrics – with confirmation

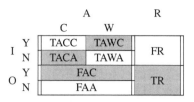

Similar to the case without confirmation, one can define an overall "good" performance metric by summing up over those individual metrics generally regarded as positive, i.e., TACA, TAWC, FAC, and TR, calling this overall metric True Confirm Total (TCT):

$$TCT = TACA + TAWC + FAC + TR. \tag{3.2}$$

3.3 Objective vs. Subjective

Both observable and hidden metrics are based on facts. It is a fact that:

- A call took 5 min and 23 s (observable).
- Four rejections were triggered (observable).
- The system hypothesized an agent request (observable).
- The caller asked for an agent (hidden).
- The caller's input was correctly understood in 90% of the cases (hidden).

There is an entirely different class of measures based on subjective judgments by human subjects that are to evaluate topics beyond observable and hidden facts such as:

- How well was the caller treated by the system (Caller Experience)?
- How well was the system treated by the caller (Caller Cooperation)?
- Was the call reason truly satisfied?

Subjective evaluation of spoken dialog systems has the advantage that it directly addresses the core questions most stakeholders, customers, consumers, project managers, voice user interface designers, quality assurance personnel, among others, have in mind when reasoning about the performance of an application. They want to see how their systems do in terms of user satisfaction and whether call reasons were truly satisfied. Objective metrics such as automation rate, handling time, or True Total are sometimes considered weak substitutes for lack of better metrics. However, the production of subjective metrics is cumbersome for four main reasons:

1. They are expensive. To produce a single subjective score, it may take 20 man minutes to listen to an entire call.

2. They are *subjective* metrics and, hence, subject to inter- and intra-subject variability (a change by 10% may be caused by the subject's mood [113]).
3. Results may not be reliable (due to 1, usually, there are only very few data points available for a given application (a couple of hundred) as compared to millions

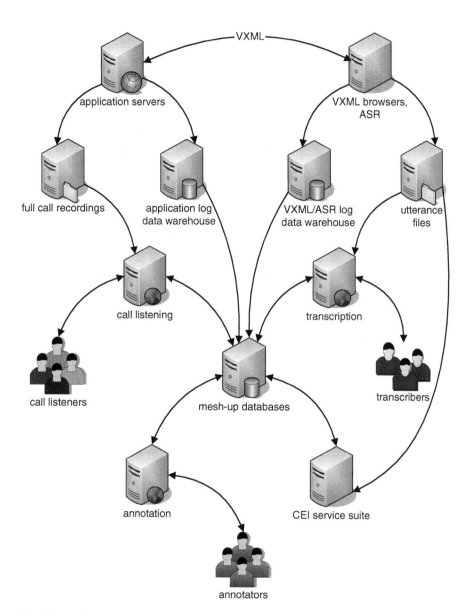

Fig. 3.2 Architecture of a deployed spoken dialog system with performance measuring infrastructure covering transcription, semantic annotation, and subjective evaluation (call listening)

in the case of freely available observable metrics; due to 2, the reliability of the individual subjective data points is somewhat weak).

4. They are not available in real time.

Consequently, the community started to investigate the possibility to predict subjective metrics based on objective ones [35, 136]. Since the correlation between objective and subjective measures can vary from application to application, the prediction algorithms need to be re-trained for new scenarios, the reason why subjective scores should be continuously collected. The constant flow of subjective evaluation is also helpful to control the accuracy of predictions and, more importantly, to catch phenomena requiring more intelligence than that of a score predictor. Examples include flaws in wording or system logic, collection of unnecessary information, or missed input utterances (speech failing to trigger the speech recognizer's endpoint detector).

3.4   Evaluation Infrastructure

Considering the call volume of deployed high-trafficked applications (see examples in Sect. 1.2), the data volume to be evaluated can be enormous. In [126], an Internet troubleshooting application with a call volume of about half a million calls per month was said to require about 1.4 TB storage in the same time frame. Considering multiple applications with even larger traffic and permanent data storage would result in data storages in the range of petabytes.

Not only does an evaluation system require a lot of storage, but the infrastructure has to be carefully engineered to account for the heterogeneous sources of data including:

- Full-duplex recordings of the whole call.
- Recordings of individual speech utterances.
- Speech recognition logs.
- Voice browser logs.
- Application logs.
- Transcriptions.
- Semantic annotations.
- Subjective ratings.

Suendermann et al. [125] describes an example of a distributed infrastructure (see Fig. 3.2) designed for this kind of large-scale evaluation, a design that was deployed in 2008 by the author and his colleagues and is being used since then.

Chapter 4
Deployed Spoken Dialog Systems' Alpha and Omega: Adaptation and Optimization

Abstract Regular tuning of spoken dialog systems is crucial to achieve maximum performance soon after the original deployment and to keep and improve the performance level during the lifetime of these systems. Often, speech recognition and understanding as well as dialog management are embedded in a continuous optimization and adaptation cycle whose details are explained in the present chapter. In addition, several techniques for quality assurance of transcription and semantic annotation as well as the statistical dialog management optimization techniques Escalator, Engager, and Contender are discussed.

Keywords Adaptation and optimization cycle • Annotation quality check • Completeness • Congruence • Consistency • Contender • Correlation • Coverage • Corpus size • Engager • Escalator • Reward • Transcription quality check

The preceding part of the present book focused on how to build deployable spoken dialog systems (Chap. 2) and how to measure their performance once being deployed (Chap. 3). Unfortunately, the results of the first performance analysis after deployment (in business lingo *post-deployment evaluation* or *post-release performance analysis*) do never ever suggest to keep the application untouched—this is somewhat counter the well-known principle on *if it ain't broke, don't fix it*. In contrast, a system not undergoing regular revisions is likely to suffer incremental performance loss until a point where the application starts producing *negative* benefits.

Negative benefits can be explained by trading off cost savings automated calls generate against costs every call produces as done in [123]. As automated calls prevent human agents from answering those calls, one can assume they saved as much as the average cost C_A induced by a human agent handling the same call type, a quantity well known to call center managers. On the flip side, automated calls produce per-minute costs C_T associated with hosting, licensing, telephony routing

D. Suendermann, *Advances in Commercial Deployment of Spoken Dialog Systems*, SpringerBriefs in Speech Technology, DOI 10.1007/978-1-4419-9610-7_4, © Springer Science+Business Media, LLC 2011

and switching maintenance, server and electricity charges, and so on. Generally, one can define a reward function for a commercial spoken dialog system as

$$R = T_A A - T \tag{4.1}$$

where A is the automation rate, T is the average handling time, and

$$T_A = \frac{C_A}{C_T}. \tag{4.2}$$

Obviously, when the automation rate falls below the critical point $\frac{T}{T_A}$, savings turn negative, and the system becomes not only useless but even hurts business.

To avoid this situation, in this chapter, a number of techniques will be discussed that can be used to continuously adapt and optimize deployed spoken dialog systems to have their performance improve over time, or, when reaching a natural saturation point, stay healthy.

4.1 Speech Recognition and Understanding

The major criticism on spoken dialog systems is their tendency to misunderstand human speech [132]. This is because speech serves as the main interface between dialog system and user and, hence, its shortcomings attract maximum attention. Problems in speech recognition and understanding cause:

- Escalations to a human upon reaching a maximum number of "speech errors" (see Sect. 3.1), hardcoded in most systems.
- Going down a wrong call flow path leading the caller into a dead end resulting in escalation to a human.
- Poor user experience making the caller hang up or ask for an agent.

Therefore, when it comes to the continuous adaptation and optimization of deployed spoken dialog systems, speech recognition and understanding are a primary topic. The following process is an example for a tuning cycle that iteratively adjusts recognition performance and is able to react to behavioral dynamics due to internal and external factors. Figure 4.1 depicts the individual steps of this cycle that are discussed in more detail below.

When a dialog system is built from scratch consisting of recognition contexts with no prior data available, system designers (voice user interface designers and speech scientists) brainstorm about what typical user utterances are to be expected in response to the contexts' system prompts. These utterances together with some optional standard robust-parsing rules (prefix, suffix, decoy) are embedded in a number of rule-based grammars as discussed in Sect. 2.3.1. Using these rule-based grammars, the initial dialog system is now deployed to production for the first time processing live traffic (on VXML application and ASR servers as shown in Fig. 3.2).

The key idea of a continuous adaptation and optimization cycle is based on the rigorous collection of speech utterances throughout all recognition contexts of the

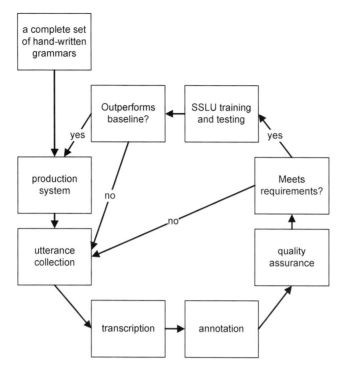

Fig. 4.1 Speech recognition and understanding continuous adaptation and optimization cycle

dialog system, a feature that is available on all major production speech recognition platforms. In order to analyze speech understanding performance of the recognition contexts of the dialog system, according to the derivations of Sect. 3.2, transcription and annotation of said utterances are required. Respective infrastructure is available in the architecture as shown in Fig. 3.2. Both transcription and annotation are primarily manual jobs but can be significantly accelerated by providing machine assistance as proposed in [122] where it is shown that a single person is able to transcribe and annotate more than 600 thousand utterances per month.

As these transcriptions and annotations are not only used for *analysis* (of recognition performance) but also for *synthesis* (of new speech recognition and understanding models, as discussed below), their quality needs to be guaranteed:

- The quality of *manual transcription* can be assured by performing regular intra- and inter-transcriber checks (i.e. assigning identical utterances either several times to the same transcriber or to different transcribers). If the test results indicate that transcription performance is suffering (transcription WER should normally be not higher than 2% [70]), the cause should be investigated and fixed.
- The derivation of *automatic transcription* as done in [122] is based on measuring manual transcription performance first and making sure that the performance of automatic transcription is not statistically significantly worse than its manual counterpart.

- To assure the quality of *manual annotation*, a number of procedures can be applied [119] including checks for:

    - *Completeness.* All utterances from a given time interval need to be completely annotated. If utterances are not yet annotated, the entire time interval should be discarded. This strict prerequisite is to make sure that the data is most representative and that there are no hidden characteristics in the non-annotated data. Take, for instance, a recognition context with only 80% annotated data whose long tail of utterances was not touched at all. If the data is used to estimate the performance of this recognition context, in the worst case, all the annotated utterances were correctly classified by the deployed speech recognition and understanding components whereas, by pure coincidence, the remaining 20% were wrong. This means the True Total on all annotated utterances was 100% whereas the actual True Total if all utterances would have been annotated would have been 80% only. Performance overestimation is a typical problem when not following the completeness check.

    - *Correlation.* Inter- and intra-annotator consistency can be evaluated similarly to what was proposed above for the quality assurance of manual transcription. Here, a useful metric is the kappa statistics [99] that expresses how strongly two sets of annotations correlate.

    - *Consistency.* Identical (or similar) utterances need to feature identical semantic annotations. Here, *similar* can mean, for example, that utterances share the same bag of words [71].

    - *Congruence.* Many utterances processed by an originally rule-based grammar in a recognition context should be covered by said grammar. Consequently, if a transcribed utterance gets successfully parsed by the grammar, it will produce the semantic class that it was designed for which can serve as ground truth, unless logical changes were applied to the semantic behavior of the recognition context. That is, most of the times a parse of a transcription is found, the parse can be directly compared to the annotation of the same utterance. They need to be identical.

    - *Coverage.* Overall speech recognition and understanding performance of recognition contexts is usually expressed by metrics such as True Total that also appreciate when out-of-grammar events get correctly rejected. Keeping this in mind, one could theoretically limit a context's scope as much as possible making almost all utterances be out of scope and then build a semantic classifier that tags every input as "out-of-scope". This way, overall understanding performance would be very high, even though almost all utterances get rejected, resulting in an entire useless scenario.

        Therefore, one needs not only to check a context's performance but also its coverage, i.e., the portion of utterances in scope of the context. This portion should generally be as large as possible (e.g. >90%) to avoid re-prompting or other actions for recovering from resulting rejections as discussed in Sect. 2.3.3 (there are exceptions to this rule since some recognition contexts are expected to feature high out-of-scope ratios because callers are likely not to say anything, as in announcement contexts, see Sect. 3.2, or produce repeated background noise, as in wait contexts, see Sect. 3.1). Coverage can be increased by:

- Broadening the scope of existing classes:
  For example, the response *I don't know* may be annotated as `help` since one may assume that providing some help could help callers understand the question better. Another frequent case are *implicit* responses such as in the following example:

  > S: So, are you connected? Please say *yes* or *no*.
  > C: I am connected.

  A generic yes/no classifier would reject the caller response as it does not clearly mean yes or no. A context-specific classifier, however, knowing the system prompt, would be able to interpret the result as confirmation and, hence, could return the class `yes`. This way, fewer user utterances would have to be rejected.
- Introducing new classes:
  For example:

  > S: How do you want to pay your bill? Please say *by credit card* or *at a payment center*.

  Unexpectedly, a high portion of callers responded

  > C: By check.

  This supposed the introduction of an additional `check` class. The introduction of new classes can be explicit (i.e., the prompt would be changed to *please say* by credit card, by check *or* at a payment center) or implicit, i.e., the prompt would remain unchanged but the user input *by check* would be handled by the application and not rejected as being out-of-scope. The latter gives the application a flavor of mixed initiative (see Sect. 2.1) whereas the former likely results in higher performance since the directing nature of the prompt helps callers phrase their choice and better understand the system's capabilities.

  – *Corpus Size.* An important aspect of performance measuring is to assure that evaluation results are of statistical significance and cover the recognition contexts' typical scenarios. Therefore, training, development, and test corpora used for the evaluation of a recognition context as well as for the production of adapted and optimized speech recognition and understanding components are expected to be of a minimum size (in the magnitude of a thousand, for example).

- *Automatic annotation* can exploit two of the techniques introduced for the quality assurance of manual annotations:

  – *Consistency.* Utterances identical (or similar) to an already annotated utterance can inherit its annotation.
  – *Congruence.* Utterances that can be parsed by the original rule-based grammar can inherit the respective parse as annotation.

Once all these quality checks prove positive, the data is split into training, development, and test data based on some heuristics. Statistical language models and classifiers are built based on the training data (see Sect. 2.3.2 for references), and parameters are tuned using the development data. Depending on the specific kind of language models and classifiers used, these parameters may include:

- Rejection threshold (see Sect. 2.3.3).
- Confirmation threshold (see Sect. 2.3.3).
- Language model/acoustic model trade-off weight [110], or
- Pruning factor [41].

The new model's performance is evaluated against the test set producing the data point TT. In order to obtain a standard of comparison, also the performance of the models currently deployed in production is measured against the test set (data point TT_0). If TT is found to be statistically significantly larger than TT_0 (e.g. based on t-test statistics [104]), the new models are registered as release candidates. If new classes were introduced, the dialog manager has to be altered to accommodate these changes. Otherwise, currently deployed models can be replaced by the release candidates at any time, including by automated deployment.

The entire above outlined cycle can be carried out in form of a 24/7 process, whereby almost all steps are fully automatic, with the exception of transcription and annotation that require some human intervention. An example of the effect of the continuous adaptation and optimization cycle on the speech understanding performance of a recognition context is shown in Fig. 4.2.

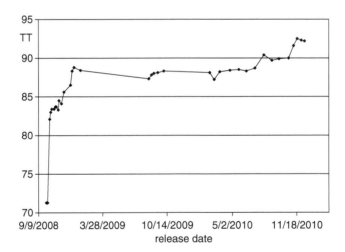

Fig. 4.2 Example of the impact of the speech recognition and understanding continuous adaptation and optimization cycle on the True Total of a recognition context. Displayed is a picture problem disambiguation context of a cable TV troubleshooting system. The question prompt reads *You can say* no picture, frozen picture, *or* poor picture quality. < 1.5 *s silence* > *Say* repeat *to hear that list again, or say* other problem *if none of these sound right*

4.2 Dialog Management

Academic research on spoken dialog systems is dominated by statistical approaches to dialog management (see also Sect. 2.4) primarily based on reinforcement learning [66] and partially observable Markov decision processes [138, 143]. The main reason for using statistics to describe a dialog manager (rather than a rule-based call flow) is that they are supposed to learn effective management strategy automatically rather than due to the intelligent architecture of a smart designer. This includes the initial design (that is often provided by an indeed rule-based simulated user), as well as adaptation to specific situations or changing environments and the long-run optimization of the application's performance. However, to the knowledge of the author, very few, if any, of these systems were ever deployed to take substantial live traffic (the systems mentioned in Sect. 2.4 processed about 60 total calls per day [34]).

Even though the deployment of fully statistical dialog managers for large-scale dialog systems seems unlikely to happen in the near future, there have been successful attempts to apply statistical adaptation and optimization techniques to deployed rule-based dialog managers three of which are discussed in this section.

4.2.1 Escalator

As suggested by the reward function expressed by (4.1), the major contributor to a commercially deployed application's *effectiveness* is its ability to automate. Another, usually less important one, is its *efficiency*, i.e., its ability to achieve its goal in as short time as possible. A significant portion of calls (often the majority) ends up non-automated. All these calls have negative rewards since the first term of (4.1) becomes zero. To increase the overall reward of an application (including automated calls), it would therefore be worthwhile to try reducing the duration of non-automated calls by escalating them as early as possible. An earlier escalation would not have a negative impact on the automation of those calls (they would remain non-automated) but a positive on the average handling time.

In order to be able to escalate non-automated calls earlier, one needs an *Escalator* (aka call outcome predictor), an algorithm that tells the dialog manager when it is confident enough that a call will not be automated. A nice property of Escalators is that their effectiveness can be evaluated offline, i.e. by applying it to log data of formerly processed sessions. This is because their presence has no impact on a given call unless they cause the call to be escalated, at which point both automation and call duration of the affected call are determined, and the call's reward can be computed. If a call is not affected, the reward remains naturally the same.

Early Escalators were implemented for AT&T's How May I Help You call router [59, 134, 135], and a first implementation using a commercial reward function as in (4.1) was described in [67]. There, the authors used two parameterizations:

M1 with $T_A = 600$ s and M2 with $T_A = 840$ s. They showed how the increase of T_A lowered the effectiveness of their technique from an average reward gain of 34.2 s (M1) to 2.4 s (M2). The parameterization assumptions, however, were far from realistic. In [116], it was shown that real-world settings of T_A are of the magnitude 5,000 s and up, i.e., significantly larger than those used in [67], so, their conclusions are not applicable to real-world-deployed systems.

Also much more recent publications on the topic such as [105], using a variety of features from all the components ASR, SLU, and dialog manager, fail to achieve a performance that would produce a positive overall reward gain. The main reason is that Escalators do not only affect calls that end up non-automated anyway but also some that would have been automated. Due to a business condition that significantly prefers automation to shortness of calls, it is much worse to classify a call that would have ended up automated as non-automated (False Accept) than to miss an early escalation due to the conviction that the call would be automated even though it was not (False Reject). That is, the *precision* (True Accept/(True Accept+False Accept)) of an Escalator must be very high to be effective.

In [123], a greedy Escalator was proposed based on discriminative training that exploits the observation that, in very complex call flows, such as the troubleshooting applications introduced in Sect. 3.1, there are branches that apparently almost never lead to automated calls. Iterating through all the activities of a call flow, one can quantify the average reward of all the calls routed through these activities. In doing so, one can produce a ranked list of the activities with their associated average rewards. Starting with the least performing activity, one can now iteratively prune the call flow at the identified worst activities, step-by-step removing more and more branches until the entire call flow has been pruned. For every step, using some test logs, one can estimate the average reward of the pruned call flow producing a function of the average reward depending on the number of pruned activities. Figure 4.3 shows an example function for an Internet troubleshooting application with the parameter settings given in Table 4.1. It shows that the original application's average reward was about 183 s whereas the version with 176 pruned nodes achieved a reward of about 196 s, all in all some 13 s gain.

4.2.2 Engager

The intent of an Escalator was to shorten the average handling time of calls whereby the overall reward according to (4.1) would be positively influenced. However, as demonstrated in Fig. 4.3, Escalators usually compromise automation rate to a certain extent limiting their overall positive impact. Taking this observation into account, the question arises whether there is a way to reduce average handling time without (negatively) impacting automation.

Considering the example of the Escalator pruning sub-trees of a given call flow, obviously, when the pruning is too aggressive, effective call branches are removed bringing the automation rate down. A different technique uses the entire original call

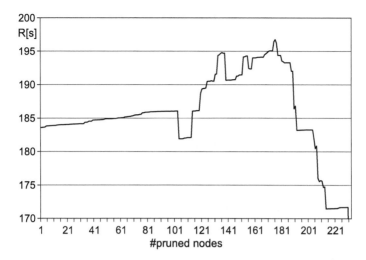

Fig. 4.3 Example of an Escalator reward function depending on the number of pruned nodes

Table 4.1 Settings for an Escalator experiment		
#calls (tokens)		45,631
#nodes (types)		847
#nodes pruned		176
T_A		5,000 s
R w/o pruning		183.5 s
R w/ pruning		196.8 s
ΔR		13.3 s

flow but revises the order in which activities are being engaged. An Engager exploits the fact that the steps carried out by the dialog manager (asking questions, querying backend devices, performing tests) may convey different levels of *informativeness*. To give an example: Imagine a dialog system is to find out which type of modem a caller has. There are three modem types:

(1) Black Ambit
(2) White Ambit
(3) Black Arris.

The voice user interface designer considers two questions to disambiguate the modem type:

(A) Is your modem black or white?
(B) Do you have an Ambit or an Arris modem?

When the answer to A is *white*, the modem is of Type 2, while the answer *Arris* to B would warrant modem Type 3, i.e., there are several cases for which only one question needs to be asked. Apparently, it depends on the prior probabilities of Types 1, 2, and 3 to decide which question should be asked first in order to minimize the average number of questions asked and, hence, minimize average handling time.

For example, take $p(1) = 0.2$, $p(2) = 0.3$, $p(3) = 0.5$. The answer to Question A is *white* with a probability of $p(2) = 0.3$, in all other cases, i.e. with the probability $p(1) + p(3) = 0.7$, Question B needs to be asked as well resulting in the average number of questions for the question order A⟶B of $p(2) + 2(p(1) + p(3)) = 1.7$.

On the other hand, with a probability of $p(3) = 0.5$, the answer to B would be *Arris*, so Question A would have to be asked with a probability $p(1) + p(2) = 0.5$. Consequently, the average number of questions for the order B⟶A is $p(3) + 2(p(1) + p(2)) = 1.5$. That is, in the example scenario, B should be asked first.

Luckily, the prior probabilities of modem types can be estimated rather reliably by looking at statistics of formerly processed calls. If there is only few or no prior call data available, rough estimates of the distributions of variables in question can often be obtained from less reliable sources such as the manufacturer of the products for which the dialog system renders support or call center managers or agents working in the same fields or market. At any rate, the Engager methodology can be useful as a tool for voice user interface designers trying to shed light on frequent uncertainties about:

- The order of activities.
- The type of questions asked (yes/no vs. small menu vs. large menu vs. open prompt).
- The useful(or -less)ness of performing certain activities at all.

In large call flows, the exhaustive consideration of every single order of activities is impossible due to its exponential growth with growing number of activities. There is, however, a number of approaches rendering Engager tractable including:

- The *negligence of activities and re-orderings* due to design and logical constraints. The following design could possibly be optimal in terms of handling time given the distribution of call reasons, however, it lacks reason (adopted from [124]):

    S: Welcome to Mewtheex. Are you calling about a red, blue, or black instrument?
    C: Uuh. I don't care.
    S: Do you need repair or do you want to buy one?
    C: Buying, I guess.
    S: Do you want to pay by credit card or check?
    C: Uuh?!
    S: And... which instrument is it about: ukulele, piccolo, or triangle? You can also say *give me a different instrument*.
    C: What the hell?! I need an Eliminator Demon Drive Double Bass Drum Pedal!

- *Considering processes.* Call flows are often subdivided into smaller units (sub-call flows, processes) whose internal activities directly relate to each other. For example, an Internet troubleshooting application can include processes for:

    - Collecting the modem type
    - Collecting the router type
    - Collecting the computer's operation system
    - Collecting the firewall brand

    and so on.

First, it does not make much sense to optimize activities across process borders for these examples, since this could result in a confusable order of things much like in the above instruments store example. So, Engager should be applied locally to the activities inside every process.

Second, processes themselves can be regarded as meta-activities whose order can be optimized by Engager. So, should the system collect the modem or the operation system first, and the like.

- *Greedy approaches.* Instead of trying every single order of activities, certain criteria of the informativeness of activities can be used to determine which question should be asked first. Informativeness measures are entropy, variance, information gain, and others [31, 68, 81]. Accordingly, Engager can be implemented as a decision tree whose nodes are the activities and whose transitions are the inputs from callers or backend devices. Standard greedy decision tree learners such as C4.5 [93] or RIPPER [23] can be used as Engagers trained on log data of a deployed application. Since decision tree learning is computationally very cheap, the Engager can dynamically change as more and more data is being collected.

4.2.3   Contender

Both approaches discussed so far, Escalator and Engager, primarily aim at increasing an application's expected reward by reducing the average handling time. As stated in Sect. 4.2.1, in most deployed systems, the main contributor to the reward is the automation rate that neither Escalator nor Engager have a clear positive impact on. As a logical consequence, the question arises which changes to the application could positively impact automation. Out of the numerous ideas flying around in voice user interface designers', system engineers', and speech scientists' minds, which ones would increase automation, and to which extent?

- Is directed dialog best in this context?
- Or open prompt?
- Open prompt given an example?
- Or two?
- Or open prompt but offering a backup menu?
- Or a yes/no question followed by an open prompt when the caller says *no*?
- What are the best examples?
- How much time should one wait before offering the backup menu?
- Which is the ideal confirmation threshold?
- What about the voice activity detection sensitivity?
- When should the recognizer time out?
- What is the best strategy following a no-match?
- Touch-tone in the first or only in the second no-match prompt?
- Or should the system go directly to the backup menu after a no-match?
- What in the case of a time-out?
- Et cetera.

Fig. 4.4 Example of a
Contender with three
alternatives

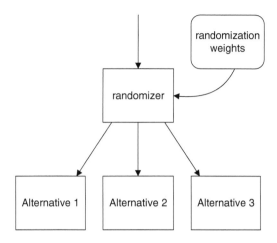

In contrast to the above discussed two techniques whose impact on handling time and automation can be approximated by analyzing log data collected on the formerly deployed system, it is practically impossible to predict the effect of arbitrary alterations such as exemplified above. Consequently, the only way to quantify their effect is to actually implement them and have them handle live traffic.

To eliminate the potential time-dependence of performance, all alternatives to a given baseline approach can be implemented in a single system, and the handled call traffic can be systematically distributed among all of them. A possible framework is the so-called Contender [121] that uses a randomization activity to decide at runtime which alternative will be used. The randomizer is parameterized by a set of weights deciding which amount of traffic will be routed to which alternative on average. Figure 4.4 displays an example Contender with three alternatives.

After collecting a certain amount of traffic for each of the alternatives, log data can be analyzed to determine how much their average rewards differ from each other. In doing so, it is essential to consider the *statistical significance* of the findings since differences may not be reliable yet when too few data points are available. In a trivial thought experiment, there are two identical alternatives, and each of them processes a single call. One of the calls happens to get successfully automated, whereas the other does not. The (trivial) automation rates of the alternatives are 100% and 0%, respectively, so, one could believe that the former is the clear winner. To overcome this dilemma, in case of a Contender with two alternatives, one may want to apply standard statistical significance tests (such as t- or z-tests [104]) whose p-value determines how likely a reward difference is by chance.

In the aforementioned example case of a Contender with two alternatives, a p-value of the null hypothesis that Alternative 1 *does not* outperform Alternative 2 leads directly to the probabilities that:

- Alternative 1 is the actual winner ($p(1) = 1 - p$).
- Alternative 2 is the actual winner ($p(2) = p$).

Fig. 4.5 Example
distribution of an
application's handling time

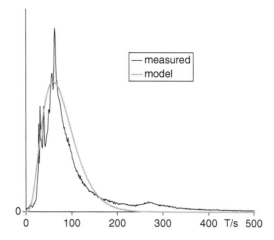

More advanced considerations into the notion of statistical significance for Contenders [121] have shown that standard significance tests may not be reliable, especially when it comes to Contenders with more than two ($I > 2$) alternatives. This is mainly because:

- They assume that the reward follows a univariate normal distribution. Considering the reward function given by (4.1), this is clearly not the case. The addend $T_A A$ follows a discrete distribution with the two values T_A (automated) and 0 (not automated) with different probabilities, and the other addend T is roughly distributed according to a Gamma distribution, see Fig. 4.5. The sum of these addends is a bivariate inverse Gamma function with the upper bound T_A.
- They require a minimum number of samples per path.
- They assume that the variances of the compared distributions are either known or equal each other.
- Most importantly, they only deliver significance estimates of pair-wise comparisons which do not straightforwardly provide general calculation rules for the winning probabilities $p(1), \ldots, p(I)$.

As recently shown (see [121]), $p(1), \ldots, p(I)$ can be estimated based on basic probabilistic relations. This requires a parametric model of the reward probability distribution to be established (see above for an example using a bivariate inverse Gamma function). For a Contender with I alternatives, an I-dimensional definite integral over the probability distribution model given reward observations for each alternative needs to be (numerically) solved to derive the order probability $p(r_i > r_j > r_k > \cdots > r_z)$, i.e. the probability that Alternative i performs better than j that performs better than k that ... that performs better than z.

Finally, to determine the winning probability of Alternative i, one needs to sum over all order probabilities headed by r_i, i.e., over $(I - 1)!$ terms altogether. For example, for a Contender with 3 alternatives, $p(1)$ can be written as

$$p(1) = p(r_1 > r_2 > r_3) + p(r_1 > r_3 > r_2). \tag{4.3}$$

The winning probability of an alternative is mainly influenced by two variables:

- The actual performance difference (the larger the difference, the faster the winning probabilities converge to 0 or 1).
- The amount of data collected for each of the alternatives (the faster data is collected, the faster the winning probabilities converge to 0 or 1).

Winning probabilities never actually become 0 or 1 but only converge to these limits. Consequently, the question arises, at which moment a decision about a final winner can be made. Similar to decisions in statistical significance analysis, a winner could be an alternative whose winning probability is greater than $1 - \alpha$ (α is the *significance level*). Typical values are $\alpha = 0.05$ and $\alpha = 0.01$.

Conventionally, in Contender experimentation, the initial weights are set to route equal traffic to all alternatives (exceptions include the case where stakeholders or other sources clearly indicate that one or more alternatives are more likely to be the winner than others). When a winner according to the above criterion has been found, the weights are changed to route the entire traffic down the winning path. There are three main issues with this approach:

1. There is a chance of approximately α that the chosen winner was not the actual one.
2. By eliminating all alternatives but the chosen winner, dynamic changes to the winning probabilities due to system alterations or external variations cannot be observed anymore.
3. Since equal amount of traffic is routed to all alternatives until a winner has been found, the cumulative performance is suboptimal.

To overcome these drawbacks, one can *use the winning probabilities as Contender weights*. In doing so, there is no need for a significance level α. By dynamically updating the weights, the convergence to 0 or 1 is acknowledged (1). By keeping a minimum amount of traffic on all alternatives (e.g. 1%), even the ones clearly suffering, the Contender keeps exploring and is able to react on changes happening to the application (2). Last but not least, it has been shown that weights based on the winning probability outperform the conventional approach in terms of cumulative application performance [121] (3).

References

1. United States Census 2000 Profile. Tech. rep., U.S. Department of Commerce, Economics and Statistics Administration, Washington, USA (2002)
2. Speech Server 2004: Product Datasheet. Tech. rep., Microsoft (2004)
3. Statistics of Communications Common Carriers: 2006/2007 Edition. Tech. rep., Federal Communications Commission, Washington, USA (2007)
4. Abello, J., Pardalos, P., Resende, M.: Handbook of Massive Data Sets. Kluwer Academic Publishers, Dordrecht, Netherlands (2002)
5. Abramowitz, M.: Handbook of Mathematical Functions with Formulas, Graphs, and Mathematical Tables. Dover, New York, USA (1964)
6. Acomb, K., Bloom, J., Dayanidhi, K., Hunter, P., Krogh, P., Levin, E., Pieraccini, R.: Technical Support Dialog Systems: Issues, Problems, and Solutions. In: Proc. of the HLT-NAACL. Rochester, USA (2007)
7. Ai, H.: User Simulation for Spoken Dialog System Development. Ph.D. thesis, University of Pittsburgh, Pittsburgh, USA (2009)
8. Allen, J., Ferguson, G., Stent, A.: An Architecture for More Realistic Conversational Systems. In: Proc. of the IUI. Santa Fe, USA (2001)
9. Alshawi, H.: The Core Language Engine. MIT Press, Cambridge, USA (1992)
10. Anton, J.: Call Center Management by the Numbers. Purdue University Press, West Lafayette, USA (1997)
11. Bacchiani, M., Beaufays, F., Schalkwyk, J., Schuster, M., Strope, B.: Deploying GOOG-411: Early Lessons in Data, Measurement, and Testing. In: Proc. of the ICASSP. Las Vegas, USA (2008)
12. Balchandran, R., Ramabhadran, L., Novak, M.: Techniques for Topic Detection Based Processing in Spoken Dialog Systems. In: Proc. of the Interspeech. Makuhari, Japan (2010)
13. Black, A., Burger, S., Langner, B., Parent, G., Eskenazi, M.: Spoken Dialog Challenge 2010. In: Proc. of the SLT. Berkeley, USA (2010)
14. Bohus, D., Rudnicky, A.: Constructing Accurate Beliefs in Spoken Dialog Systems. In: Proc. of the ASRU. San Juan, Puerto Rico (2005)
15. ten Bosch, L., Oostdijk, N., Boves, L.: On Temporal Aspects of Turn Taking in Conversational Dialogues. Speech Communication **47**(1/2) (2005)
16. Boulanger, D., Bruynooghe, M.: Using Call/Exit Analysis for Logic Program Transformation. In: Proc. of the LOBSTR. Pisa, Italy (1994)
17. Boyce, S.: User interface design for natural language systems: From research to reality. In: D. Gardner-Bonneau, H. Blanchard (eds.) Human Factors and Voice Interactive Systems. Springer, New York, USA (2008)
18. Brants, T., Franz, A.: Web 1T 5-Gram Corpus Version 1.1. Tech. rep., Google Research (2006)

D. Suendermann, *Advances in Commercial Deployment of Spoken Dialog Systems*,
SpringerBriefs in Speech Technology, DOI 10.1007/978-1-4419-9610-7,
© Springer Science+Business Media, LLC 2011

19. Burke, D.: Speech Processing for IP Networks: Media Resource Control Protocol (MRCP). Wiley, New York, USA (2007)

20. Chai, J., Horvath, V., Nicolov, N., Stys-Budzikowska, M., Kambhatla, N., Zadrozny, W.: Natural Language Sales Assistant – A Web-Based Dialog System for Online Sales. In: Proc. of the Conference on Innovative Applications of Arificial Intelligence. Seattle, USA (2001)

21. Chandramohan, S., Geist, M., Pietquin, O.: Optimizing Spoken Dialogue Management with Fitted Value Iteration. In: Proc. of the Interspeech. Makuhari, Japan (2010)

22. Cohen, M., Giangola, J., Balogh, J.: Voice User Interface Design. Addison-Wesley, Redwood City, USA (2004)

23. Cohen, W.: Fast Effective Rule Induction. In: Proc. of the International Conference on Machine Learning. Lake Tahoe, USA (1995)

24. Cox, R., Kamm, C., Rabiner, L., Schroeter, J., Wilpon, J.: Speech and Language Processing for Next-Millennium Communications Services. Proc. of the IEEE **88**(8) (2000)

25. Dahl, D.: Practical Spoken Dialog Systems. Springer, New York, USA (2006)

26. Dale, R., Reiter, E.: Building Natural Language Generation Systems. Cambridge University Press, Cambridge, UK (2000)

27. Davis, K., Biddulph, R., Balashek, S.: Automatic Recognition of Spoken Digits. Journal of the Acoustical Society of America **24**(6) (1952)

28. Devillers, L.: Evaluation of Dialog Strategies for a Tourist Information Retrieval System. In: Proc. of the ICSLP. Sydney, Australia (1998)

29. Dinarelli, M.: Spoken Language Understanding: From Spoken Utterances to Semantic Structures. Ph.D. thesis, University of Trento, Povo, Italy (2010)

30. Dybkjær, H., Dybkjær, L.: Modeling Complex Spoken Dialog. Computer Journal **37**(8) (2004)

31. Ebrahimi, N., Maasoumi, E., Soofi, E.: Measuring informativeness of data by entropy and variance. In: D. Slottje (ed.) Essays in Honor of Camilo Dagum. Physica, Heidelberg, Germany (1999)

32. ECMA: Standard ECMA-262 ECMAScript Language Specification. http://www.ecma-international.org/publications/standards/Ecma-262.htm (1999)

33. Egges, A., Nijholt, A., op den Akker, H.: Dialogs with BDP Agents in Virtual Environments. In: Proc. of the IJCAI. Seattle, USA (2001)

34. Eskenazi, M., Black, A., Raux, A., Langner, B.: Let's Go Lab: A Platform for Evaluation of Spoken Dialog Systems with Real World Users. In: Proc. of the Interspeech. Brisbane, Australia (2008)

35. Evanini, K., Hunter, P., Liscombe, J., Suendermann, D., Dayanidhi, K., Pieraccini:, R.: Caller Experience: A Method for Evaluating Dialog Systems and Its Automatic Prediction. In: Proc. of the SLT. Goa, India (2008)

36. Evanini, K., Suendermann, D., Pieraccini, R.: Call Classification for Automated Troubleshooting on Large Corpora. In: Proc. of the ASRU. Kyoto, Japan (2007)

37. Evermann, G., Chan, H., Gales, M., Jia, B., Mrva, D., Woodland, P., Yu, K.: Training LVCSR Systems on Thousands of Hours of Data. In: Proc. of the ICASSP. Philadelphia, USA (2005)

38. di Fabbrizio, G., Tur, G., Hakkani-Tür, D.: Bootstrapping Spoken Dialog Systems with Data Reuse. In: Proc. of the SIGdial Workshop on Discourse and Dialogue. Cambridge, USA (2004)

39. Galley, M., Fosler-Lussier, E., Potamianos, A.: Hybrid Natural Language Generation for Spoken Dialogue Systems. In: Proc. of the Eurospeech. Aalborg, Denmark (2001)

40. Giraudo, E., Baggia, P.: EVALITA 2009: Loquendo Spoken Dialog System. In: Proc. of the Conference of the Italian Association for Artificial Intelligence. Reggio Emilia, Italy (2004)

41. Goodman, J., Gao, J.: Language Model Size Reduction by Pruning and Clustering. In: Proc. of the ICSLP. Beijing, China (2000)

42. Goodwin, C.: Conversational Organization: Interaction Between Speakers and Hearers. Academic Press, New York, USA (1981)

43. Gorin, A., Riccardi, G., Wright, J.: How May I Help You? Speech Communication **23**(1/2) (1997)

44. Gray, J.: What Next? A Dozen Information-Technology Research Goals. Journal of the ACM **50**(1) (2003)

45. Hirschman, L., Seneff, S., Goodine, D., Phillips, M.: Integrating Syntax and Semantics into Spoken Language Understanding. In: Proc. of the HLT. Pacific Grove, USA (1991)

46. Huang, J., Gao, J., Miao, J., Li, X., Wang, K., Behr, F.: Exploring Web Scale Language Models for Search Query Processing. In: Proc. of the WWW Conference. Raleigh, USA (2010)

47. Hunt, A., McGlashan, S.: Speech Recognition Grammar Specification Version 1.0. W3C Recommendation. http://www.w3.org/TR/2004/REC-speech-grammar-20040316 (2004)

48. Hurtado, L., Griol, D., Sanchis, E., Segarra, E.: A statistical user simulation technique for the improvement of a spoken dialog system. In: L. Rueda, D. Mery, J. Kittler (eds.) Progress in Pattern Recognition, Image Analysis and Applications. Springer, New York, USA (2007)

49. Hurtado, L., Planells, J., Segarra, E., Sanchis, E., Griol, D.: A Stochastic Finite-State Transducer Approach to Spoken Dialog Management. In: Proc. of the Interspeech. Makuhari, Japan (2010)

50. Jefferson, G.: Preliminary notes on a possible metric which provides for a 'standard maximum' silence of approximately one second in conversation. In: D. Roger, P. Bull (eds.) Conversation: An Interdisciplinary Perspective. Multilingual Matters, Clevedon, UK (2007)

51. Johnston, A.: SIP: Understanding the Session Initiation Protocol. Artech House, Norwood, USA (2004)

52. Jokinen, K., McTear, M.: Spoken Dialogue Systems. Morgan & Claypool, San Rafael, USA (2010)

53. Jonsdottir, G., Gratch, J., Fast, E., Thórisson, K.: Fluid Semantic Back-Channel Feedback in Dialogue: Challenges and Progress. In: Proc. of the IVA. Paris, France (2007)

54. Juang, B., Furui, S.: Automatic Recognition and Understanding of Spoken Language–A First Step toward Natural Human-Machine Communication. Proc. of the IEEE **88**(8) (2000)

55. Jurčíček, F., Thomson, B., Keizer, S., Mairesse, F., Gašić, M., Yu, K., Young, S.: Natural Belief-Critic: A Reinforcement Algorithm for Parameter Estimation in Statistical Spoken Dialogue Systems. In: Proc. of the Interspeech. Makuhari, Japan (2010)

56. Kaelbling, L., Littman, M., Moore, A.: Reinforcement Learning: A Survey. Journal of Artificial Intelligence Research **4** (1996)

57. King, S., Karaiskos, V.: The Blizzard Challenge 2010. In: Blizzard Challenge Workshop. Kansai Science City, Japan (2010)

58. Kuhn, R., de Mori, R.: The Application of Semantic Classification Trees to Natural Language Understanding. IEEE Trans. on Pattern Analysis and Machine Intelligence **17**(5) (1995)

59. Langkilde, I., Walker, M., Wright, J., Gorin, A., Litman, D.: Automatic Prediction of Problematic Human-Computer Dialogues in How May I Help You? In: Proc. of the ASRU. Keystone, USA (1999)

60. Langner, B., Vogel, S., Black, A.: Evaluating a Dialog Language Generation System: Comparing the MOUNTAIN System to Other NLG Approaches. In: Proc. of the Interspeech. Makuhari, Japan (2010)

61. Larson, J.: Introduction and Overview of W3C Speech Interface Framework. W3C Working Draft. http://www.w3.org/TR/voice-intro (2000)

62. Lefèvre, F., Mairesse, F., Young, S.: Cross-Lingual Spoken Language Understanding from Unaligned Data Using Discriminative Classification Models and Machine Translation. In: Proc. of the Interspeech. Makuhari, Japan (2010)

63. Lésperance, Y., Levesque, H., Lin, F., Marcu, D., Reiter, R., Scherl, R.: Foundations of a Logical Approach to Agent Programming. In: Proc. of the IJCAI. Montréal, Canada (1995)

64. Levenshtein, V.: Binary Codes Capable of Correcting Deletions, Insertions, and Reversals. Soviet Physics Doklady **10** (1966)

65. Levin, E., Narayanan, S., Pieraccini, R., Biatov, K., Bocchieri, E., di Fabbrizio, G., Eckert, W., Lee, S., Pokrovsky, A., Rahim, M., Ruscitti, P., Walker, M.: The AT&T-DARPA Communicator Mixed-Initiative Spoken Dialog System. In: Proc. of the ICSLP. Beijing, China (2000)

66. Levin, E., Pieraccini, R.: A Stochastic Model of Computer-Human Interaction for Learning Dialogue Strategies. In: Proc. of the Eurospeech. Rhodes, Greece (1997)

67. Levin, E., Pieraccini, R.: Value-Based Optimal Decision for Dialog Systems. In: Proc. of the SLT. Palm Beach, Aruba (2006)

68. Lippi, M., Jaeger, M., Frasconi, P., Passerini, A.: Relational Information Gain. In: Proc. of the ILP. Leuven, Belgium (2009)

69. López-Cózar, R., Griol, D.: New Technique to Enhance the Performance of Spoken Dialogue Systems Based on Dialogue States-Dependent Language Models and Grammatical Rules. In: Proc. of the Interspeech. Makuhari, Japan (2010)

70. Marge, M., Banerjee, S., Rudnicky, A.: Using the Amazon Mechanical Turk for Transcription of Spoken Language. In: Proc. of the ICASSP. Dallas, USA (2010)

71. McCallum, A., Nigam, K.: A Comparison of Event Models for Naive Bayes Text Classification. In: Proc. of the AAAI Workshop on Learning for Text Categorization. Madison, USA (1998)

72. McGlashan, S., Burnett, D., Carter, J., Danielsen, P., Ferrans, J., Hunt, A., Lucas, B., Porter, B., Rehor, K., Tryphonas, S.: VoiceXML 2.0. W3C Recommendation. http://www.w3.org/TR/2004/REC-voicexml20-20040316 (2004)

73. McInnes, F., Nairn, I., Attwater, D., Jack, M.: Effects of Prompt Style on User Responses to an Automated Banking Service Using Word-Spotting. BT Technology 17(1) (1999)

74. McTear, M.: Spoken Dialogue Technology. Springer, New York, USA (2004)

75. McTear, M.: Spoken Language Understanding for Conversational Dialog Systems. In: Proc. of the SLT. Palm Beach, Aruba (2006)

76. Melin, H., Sandell, A., Ihse, M.: CTT-Bank: A Speech Controlled Telephone Banking System – An Initial Evaluation. Tech. rep., KTH, Stockholm, Sweden (2001)

77. Messerli, E.: Proof of a Convexity Property of the Erlang B Formula. Bell System Technical Journal 51(4) (1972)

78. Miller, S., Bobrow, R., Ingria, R., Schwartz, R.: Hidden Understanding Models of Natural Language. In: Proc. of the ACL. Las Cruces, USA (1994)

79. Minker, W., Bennacef, S.: Speech and Human-Machine Dialog. Springer, New York, USA (2004)

80. Minker, W., Lee, G., Nakamura, S., Mariani, J.: Spoken Dialogue Systems Technology and Design. Springer, New York, USA (2011)

81. Mitchell, T.: Machine Learning. McGraw Hill, New York, USA (1997)

82. Moore, R.: Presence: A Human-Inspired Architecture for Speech-Based Human-Machine Interaction. IEEE Trans. on Computers 56(9) (2007)

83. de Mori, R.: Spoken Dialogue with Computers. Academic Press, San Diego, USA (1998)

84. Nöth, E., de Mori, R., Fischer, J., Gebhard, A., Harbeck, S., Kompe, R., Kuhn, R., Niemann, H., Mast, M.: An Integrated Model of Acoustics and Language Using Semantic Classification Trees. In: Proc. of the ICASSP. Atlanta, USA (1996)

85. Paiva, A., Prada, R., Picard, R.: Affective Computing and Intelligent Interaction. Springer, New York, USA (2007)

86. Pieraccini, R., Caskey, S., Dayanidhi, K., Carpenter, B., Phillips, M.: ETUDE, a Recursive Dialog Manager with Embedded User Interface Patterns. In: Proc. of the ASRU. Madonna di Campiglio, Italy (2001)

87. Pieraccini, R., Levin, E.: Stochastic Representation of Semantic Structure for Speech Understanding. In: Proc. of the Eurospeech. Genova, Italy (1991)

88. Pieraccini, R., Lubensky, D.: Spoken language communication with machines: The long and winding road from research to business. In: M. Ali, F. Esposito (eds.) Innovations in Applied Artificial Intelligence. Springer, New York, USA (2005)

89. Potamianos, A., Ammicht, E., Kuo, J.: Dialogue Management in the Bell Labs Communicator System. In: Proc. of the ICSLP. Beijing, China (2000)

90. Price, P.: Evaluation of Spoken Language Systems: The ATIS Domain. In: Proc. of the Workshop on Speech and Natural Language. Hidden Valley, USA (1990)

91. Putze, F., Schultz, T.: Utterance Selection for Speech Acts in a Cognitive Tourguide Scenario. In: Proc. of the Interspeech. Makuhari, Japan (2010)

92. Quarteroni, S., González, M., Riccardi, G., Varges, S.: Combining User Intention and Error Modeling for Statistical Dialog Simulators. In: Proc. of the Interspeech. Makuhari, Japan (2010)

93. Quinlan, J.: C4.5: Programs for Machine Learning. Morgan Kaufmann, San Francisco, USA (1993)

94. Rambow, O., Bangalore, S., Walker, M.: Natural Language Generation in Dialog Systems. In: Proc. of the HLT. San Diego, USA (2001)

95. Raux, A., Bohus, D., Langner, B., Black, A., Eskenazi, M.: Doing Research on a Deployed Spoken Dialogue System: One Year of Let's Go! Experience. In: Proc. of the Interspeech. Pittsburgh, USA (2006)

96. Raux, A., Langner, B., Black, A., Eskenazi, M.: LET'S GO: Improving Spoken Dialog Systems for the Elderly and Non-Native. In: Proc. of the Eurospeech. Geneva, Switzerland (2003)

97. Raux, A., Mehta, N., Ramachandran, D., Gupta, R.: Dynamic Language Modeling Using Bayesian Networks for Spoken Dialog Systems. In: Proc. of the Interspeech. Makuhari, Japan (2010)

98. Rodgers, J., Nicewander, W.: Thirteen Ways to Look at the Correlation Coefficient. The American Statistician **42**(1) (1988)

99. Rosenberg, A., Binkowski, E.: Augmenting the Kappa Statistic to Determine Interannotator Reliability for Multiply Labeled Data Points. In: Proc. of the HLT/NAACL. Boston, USA (2004)

100. Rosenfeld, R.: Two Decades of Statistical Language Modeling: Where Do We Go from Here? Proc. of the IEEE **88**(8) (2000)

101. Rotaru, M.: Applications of Discourse Structure for Spoken Dialogue Systems. Ph.D. thesis, University of Pittsburgh, Pittsburgh, USA (2008)

102. Rudnicky, A., Xu, W.: An Agenda-Based Dialog Management Architecture for Spoken Language Systems. In: Proc. of the ASRU. Keystone, USA (1999)

103. Sadek, M., Bretier, P., Panaget, F.: Artimis: Natural Dialogue Meets Rational Agency. In: Proc. of the IJCAI. Nagoya, Japan (1997)

104. Schervish, M.: Theory of Statistics. Springer, New York, USA (1995)

105. Schmitt, A., Scholz, M., Minker, W., Liscombe, J., Suendermann, D.: Is it Possible to Predict Task Completion in Automated Troubleshooters? In: Proc. of the Interspeech. Makuhari, Japan (2010)

106. Shanmugham, S., Monaco, P., Eberman, B.: A Media Resource Control Protocol (MRCP): Internet Society Request for Comments. http://tools.ietf.org/html/rfc4463 (2006)

107. Singh, K., Park, D.: Economical Global Access to a VoiceXML Gateway Using Open Source Technologies. In: Proc. of the Coling. Manchester, UK (2008)

108. Souvignier, B., Kellner, A., Rueber, B., Schramm, H., Seide, F.: The Thoughtful Elephant: Strategies for Spoken Dialog Systems. IEEE Trans. on Speech and Audio Processing **8**(1) (2000)

109. Srinivasan, S., Brown, E.: Is Speech Recognition Becoming Mainstream? Computer Journal **35**(4) (2002)

110. Stemmer, G., Zeißler, V., Nöth, E., Niemann, H.: Towards a Dynamic Adjustment of the Language Weight. In: Proc. of the TSD. Zelezna Ruda, Czech Republic (2001)

111. Stent, A.: Dialogue Systems as Conversational Partners: Applying Conversation Acts Theory to Natural Language Generation for Task-Oriented Mixed-Initiative Spoken Dialogue. Ph.D. thesis, University of Rochester, Rochester, USA (2001)

112. Stent, A., Stenchikova, S., Marge, M.: Dialog Systems for Surveys: The Rate-a-Course System. In: Proc. of the SLT. Palm Beach, Aruba (2006)

113. Suendermann, D.: Text-Independent Voice Conversion. Ph.D. thesis, Bundeswehr University Munich, Munich, Germany (2008)

114. Suendermann, D., Hoege, H., Black, A.: Challenges in speech synthesis. In: F. Chen, K. Jokinen (eds.) Speech Technology: Theory and Applications. Springer, New York, USA (2010)

115. Suendermann, D., Hunter, P., Pieraccini, R.: Call Classification with Hundreds of Classes and Hundred Thousands of Training Utterances ... and No Target Domain Data. In: Proc. of the PIT. Kloster Irsee, Germany (2008)

116. Suendermann, D., Liscombe, J., Bloom, J., Li, G., Pieraccini, R.: Deploying Contender: Early Lessons in Data, Measurement, and Testing of Multiple Call Flow Decisions. In: Proc. of the HCI. Washington, USA (2011)

117. Suendermann, D., Liscombe, J., Dayanidhi, K., Pieraccini, R.: A Handsome Set of Metrics to Measure Utterance Classification Performance in Spoken Dialog Systems. In: Proc. of the SIGdial Workshop on Discourse and Dialogue. London, UK (2009)

118. Suendermann, D., Liscombe, J., Dayanidhi, K., Pieraccini, R.: Localization of Speech Recognition in Spoken Dialog Systems: How Machine Translation Can Make Our Lives Easier. In: Proc. of the Interspeech. Brighton, UK (2009)

119. Suendermann, D., Liscombe, J., Evanini, K., Dayanidhi, K., Pieraccini, R.: C^5. In: Proc. of the SLT. Goa, India (2008)

120. Suendermann, D., Liscombe, J., Evanini, K., Dayanidhi, K., Pieraccini, R.: From Rule-Based to Statistical Grammars: Continuous Improvement of Large-Scale Spoken Dialog Systems. In: Proc. of the ICASSP. Taipei, Taiwan (2009)

121. Suendermann, D., Liscombe, J., Pieraccini, R.: Contender. In: Proc. of the SLT. Berkeley, USA (2010)

122. Suendermann, D., Liscombe, J., Pieraccini, R.: How to Drink from a Fire Hose: One Person Can Annoscribe 693 Thousand Utterances in One Month. In: Proc. of the SIGdial Workshop on Discourse and Dialogue. Tokyo, Japan (2010)

123. Suendermann, D., Liscombe, J., Pieraccini, R.: Minimally Invasive Surgery for Spoken Dialog Systems. In: Proc. of the Interspeech. Makuhari, Japan (2010)

124. Suendermann, D., Liscombe, J., Pieraccini, R.: Optimize the Obvious: Automatic Call Flow Generation. In: Proc. of the ICASSP. Dallas, USA (2010)

125. Suendermann, D., Liscombe, J., Pieraccini, R., Evanini, K.: 'How am I Doing?' A New Framework to Effectively Measure the Performance of Automated Customer Care Contact Centers. In: A. Neustein (ed.) Advances in Speech Recognition: Mobile Environments, Call Centers and Clinics. Springer, New York, USA (2010)

126. Suendermann, D., Pieraccini, R.: SLU in commercial and research spoken dialogue systems. In: G. Tur, R. de Mori (eds.) Spoken Language Understanding. Wiley, New York, USA (2011)

127. Suhm, B., Peterson, P.: A Data-Driven Methodology for Evaluating and Optimizing Call Center IVRs. Speech Technology **5**(1) (2002)

128. Sun Microsystems: Java Speech Grammar Format Specification Version 1.0. http://java.sun.com/products/java-media/speech/forDevelopers/JSGF/ (1998)

129. Syrdal, A., Kim, Y.J.: Dialog Speech Acts and Prosody: Considerations for TTS . In: Proc. of the Speech Prosody. Campinas, Brazil (2008)

130. Thomson, B., Yu, K., Keizer, S., Gašić, M., Jurčíček, F., Mairesse, F., Young, S.: Bayesian Dialogue System for the Let's Go Spoken Dialogue Challenge. In: Proc. of the SLT. Berkeley, USA (2010)

131. Thórisson, K.: Natural turn-taking needs no manual: Computational theory and model, from perception to action. In: I. Granström, D. House (eds.) Multimodality in Language and Speech Systems. Kluwer Academic Publishers, Dordrecht, Netherlands (2002)

132. Tomko, S.: Improving User Interaction with Spoken Dialog Systems via Shaping. Ph.D. thesis, Carnegie Mellon University, Pittsburgh, USA (2006)

133. Walker, D.: Speech Understanding through Syntactic and Semantic Analysis. IEEE Trans. on Computers **25**(4) (1976)

134. Walker, M., Langkilde, I., Wright, J., Gorin, A., Litman, D.: Learning to Predict Problematic Situations in a Spoken Dialogue System: Experiments with HMIHY?. In: Proc. of the NAACL. Seattle, USA (2000)

135. Walker, M., Langkilde-Geary, I., Hastie, H., Wright, J., Gorin, A.: Automatically Training a Problematic Dialog Predictor for the HMIHY Spoken Dialogue System. Journal of Artificial Intelligence Research **16** (2002)
136. Walker, M., Litman, D., Kamm, C.: Evaluating Spoken Dialogue Agents with PARADISE: Two Case Studies. Computer Speech and Language **12**(3) (1998)
137. Wang, Y.: Semantic frame based spoken language understanding. In: G. Tur, R. de Mori (eds.) Spoken Language Understanding. Wiley, New York, USA (2011)
138. Williams, J.: Partially Observable Markov Decision Processes for Spoken Dialogue Management. Ph.D. thesis, Cambridge University, Cambridge, UK (2006)
139. Williams, J.: Exploiting the ASR N-Best by Tracking Multiple Dialog State Hypotheses. In: Proc. of the Interspeech. Brisbane, Australia (2008)
140. Williams, J., Arizmendi, I., Conkie, A.: Demonstration of AT&T "LET'S GO": A Production-Grade Statistical Spoken Dialog System. In: Proc. of the SLT. Berkeley, USA (2010)
141. Williams, J., Witt, S.: A Comparison of Dialog Strategies for Call Routing. Speech Technology **7**(1) (2004)
142. Wilpon, J., Roe, D.: AT&T Telephone Network Applications of Speech Recognition. In: Proc. of the COST232 Workshop. Rome, Italy (1992)
143. Young, S.: Talking to Machines (Statistically Speaking). In: Proc. of the ICSLP. Denver, USA (2002)
144. Young, S., Schatzmann, J., Weilhammer, K., Ye, H.: The Hidden Information State Approach to Dialog Management. In: Proc. of the ICASSP. Hawaii, USA (2007)
145. Yuk, D., Flanagan, J.: Telephone Speech Recognition Using Neural Networks and Hidden Markov Models. In: Proc. of the ICASSP. Phoenix, USA (1999)